Test Yourself

Anatomy and Physiology

Elward K. Alford, Ph.D.
Volunteer State College
Gallatin, TN

Leane Roffey, Ph.D.
NeuroMagnetic Systems
San Antonio, TX

Contributing Editors

Susan Maskel, Ph.D.
Department of Bioological and Environmental Science
Western Connecticut State university
Danbury, CT

Mara Lauterbach, Ph.D.
Department of Biological Sciences
Towson State University
Towson, MD

Anne Anastasia, Ed.D.
Department of Biological Sciences
Fairleigh Dickinson University
Teaneck, NJ

NTC LEARNINGWORKS
NTC/Contemporary Publishing Group

Library of Congress Cataloging-in-Publication Data

Alford, Elward K.
Anatomy and physiology / Elward K. Alford, Leane Roffey ; contributing editors, Susan Maskel, Mara Lauterbach, Anne Anastasia.
p. cm. — (Test yourself)
ISBN 0-8442-2380-8
1. Human physiology—Examinations, questions, etc. I. Roffey, Leane. II. Maskel, Susan. III. Lauterbach, Mara. IV. Anastasia, Anne. V. Title. VI. Series: Test yourself (Lincolnwood, Ill.)
[DNLM: 1. Anatomy examination questions. 2. Physiology examination questions. QS 18.2 A389a 1998]
QP40.A54 1998
612′.0076—dc21
DNLM/DLC
for Library of Congress 98-13315
CIP

A *Test Yourself Books, Inc.* **Project**

Published by NTC LearningWorks
A division of NTC/Contemporary Publishing Group, Inc.
4255 West Touhy Avenue, Lincolnwood (Chicago), Illinois 60646-1975 U.S.A.

Printed in the United States of America
International Standard Book Number: 0-8442-2380-8
18 17 16 15 14 13 12 11 10 9 8 7 6 5 4 3 2 1

Contents

Preface

The study of human anatomy and physiology is an exercise that demands not only memorization of detail, but also a comprehension of the interaction of the components of the human organism. While anatomy is the study of the structures of the body, physiology is a study of the function of these structures. The delineation of separation of structure and function is largely an arbitrary construct.

In order to learn human anatomy and physiology, it is necessary to examine structure and function together, to identify the relationship between structure and function, and to understand how anatomical constraints effect physiology, as well as how physiological requirements drive anatomical development.

This study guide is designed to pose questions that not only test and train your knowledge of the details of anatomy and physiology, but to also test and train your knowledge of the interaction of anatomy and physiology in order to foster a deeper comprehension of the subject matter.

Elward K. Alford, Ph.D.
Leane Roffey, Ph.D.

How to Use This Book

This "Test Yourself" book is part of a unique series designed to help you improve your test scores on almost any type of examination you will face. Too often, you will study for a test—quiz, midterm, or final—and come away with a score that is lower than anticipated. Why? Because there is no way for you to really know how much you understand a topic until you've taken a test. The *purpose* of the test, after all, is to test your complete understanding of the material.

The "Test Yourself" series offers you a way to improve your scores and to actually test your knowledge at the time you use this book. Consider each chapter a diagnostic pretest in a specific topic. Answer the questions, check your answers, and then give yourself a grade. Then, and only then, will you know where your strengths and, more important, weaknesses are. Once these areas are identified, you can strategically focus your study on those topics that need additional work.

Each book in this series presents a specific subject in an organized manner, and although each "Test Yourself" chapter may not correspond to exactly the same chapter in your textbook, you should have little difficulty in locating the specific topic you are studying. Written by educators in the field, each book is designed to correspond, as much as possible, to the leading textbooks. This means that you can feel confident in using this book, and that regardless of your textbook, professor, or school, you will be much better prepared for anything you will encounter on your test.

Each chapter has four parts:

Brief Yourself. All chapters contain a brief overview of the topic that is intended to give you a more thorough understanding of the material with which you need to be familiar. Sometimes this information is presented at the beginning of the chapter, and sometimes it flows throughout the chapter, to review your understanding of various *units* within the chapter.

Test Yourself. Each chapter covers a specific topic corresponding to one that you will find in your textbook. Answer the questions, either on a separate page or directly in the book, if there is room.

Check Yourself. Check your answers. Every question is fully answered and explained. These answers will be the key to your increased understanding. If you answered the question incorrectly, read the explanations to *learn* and *understand* the material. You will note that at the end of every answer you will be referred to a specific subtopic within that chapter, so you can focus your studying and prepare more efficiently.

Grade Yourself. At the end of each chapter is a self-diagnostic key. By indicating on this form the numbers of those questions you answered incorrectly, you will have a clear picture of your weak areas.

There are no secrets to test success. Only good preparation can guarantee higher grades. By utilizing this "Test Yourself" book, you will have a better chance of improving your scores and understanding the subject more fully.

Introduction to the Human Body

Brief Yourself

The study of the human body is a fascinating endeavor. Students of human anatomy and physiology will acquire a more complete understanding of their own bodies, both in its structure and function. Such basic knowledge allows a context for the wealth of information produced by today's research in such diverse fields as nutrition, health, genetics, molecular biology, and other disiplines that attempt to provide new insight into the physical human being.

As with most fields of study, a familiarity with basic terminology and organization of the subject area is necessary. Just as reading a map requires a pre-existing knowledge of directions such as north, south, east, and west, the study of the human body requires a knowledge of anatomical terminology to provide a standard for the position and description of the structures of the body.

In addition, an understanding of the organization and complexity of a living organism is needed. A comprehensive introduction to the human body will identify the basic requirements and functional processes common to all living organisms and introduce the various systems of the body responsible for these processes.

Test Yourself

1. Define and contrast anatomy and physiology.

2. Anatomy and physiology contain many subdivisions. Briefly describe each of the subdivisions listed below:

   a. Systemic anatomy

   b. Developmental anatomy

   c. Pathological anatomy

   d. Exercise physiology

   e. Endocrinology

   f. Immunology

3. The levels of structural organization of the human body are often described as a hierarchy of increasing complexity of matter. List these levels in the order of their increasing complexity.

4. Identify the four tissues of the human body.

5. Define an organ.

Questions 6–10 are matching. Match the following organ systems to their components:

b	6. Integumentary system	a. Nasal cavity, pharynx, larynx, trachea, bronchi, lungs
d	7. Skeletal sytem	b. Skin, hair, nails, sweat and sebaceous glands
a	8. Respiratory system	c. Kidneys, ureters, urinary bladder, urethra
c	9. Urinary system	d. Bones, cartilage, joints
e	10. Digestive system	e. Mouth, esophagus, stomach, small intestine, large intestine, rectum, anus

11. Although life has proven to be very difficult to define, certain functional characteristics are common to all living things. List these common life processes.

12. Define homeostasis. What is its significance to maintaining life?

13. What are the physical requirements for human life?

14. What is physiological stress?

15. Why is internal communication necessary for homeostasis and what systems are specialized to do this task?

16. Most homeostatic control mechanism are feedback systems (loops). What are the three components of a feedback loop?

17. Give an example of a positive feedback system in human physiology.

18. Give an example of a negative feedback system in human physiology.

19. Describe the anatomical position.

20. What is the significance of identifying a standard anatomical position?

Questions 21–30 are matching. Match the anatomical direction with its description.

c	21. Distal	a. Away from the surface of the body
j	22. Superior (cephalic)	b. Toward the surface of the body
a	23. Deep (profound)	c. Farther from the attachment of a limb to the trunk
f	24. Medial	d. Nearer to the attachment of a limb to the trunk
d	25. Proximal	e. Farther from the midline of the body (or structure)
i	26. Inferior (caudal)	f. Nearer to the midline of the body (or structure)
b	27. Superficial	g. Nearer to the back of the body
e	28. Lateral	h. Nearer to the front of the body
g	29. Posterior (dorsal)	i. Away from the head (or toward the lower part of a structure)
h	30. Anterior (ventral)	j. Toward the head (or toward the upper part of a structure)

31. Examine the figure below. Identify the planes which divide the body.

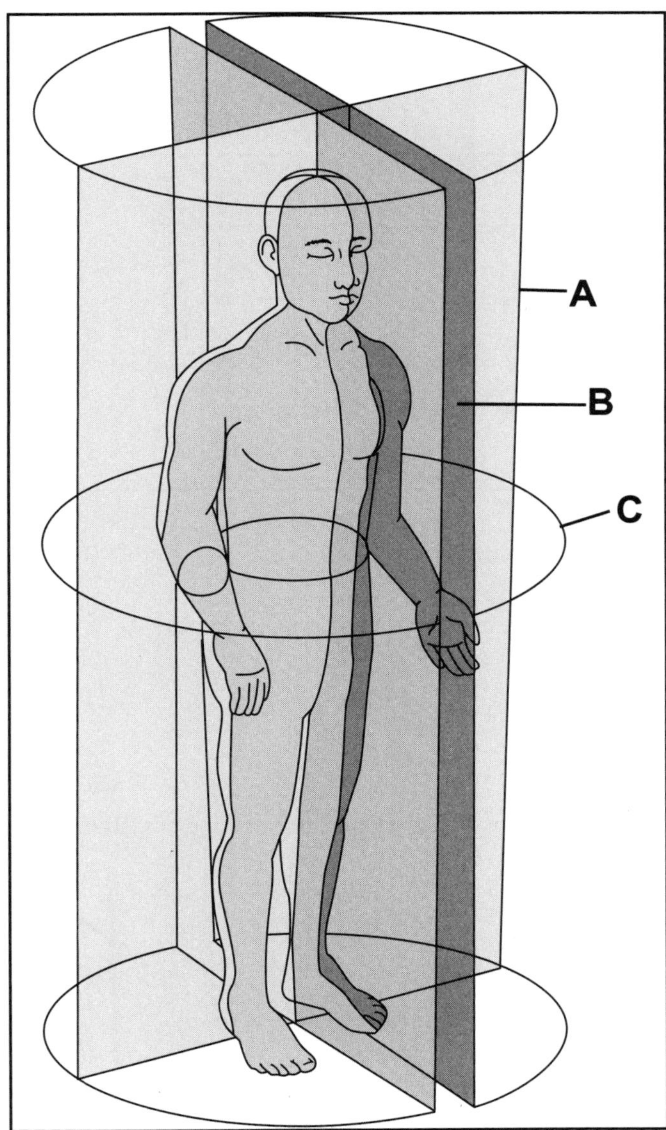

Fig. 1-1

a. frontal

b. sagital

c. transverse

32. What are the two principal cavities of the human body? What are their subdivisions?

33. The abdominopelvic cavity is divided into nine regions. Identify these regions on the figure below.

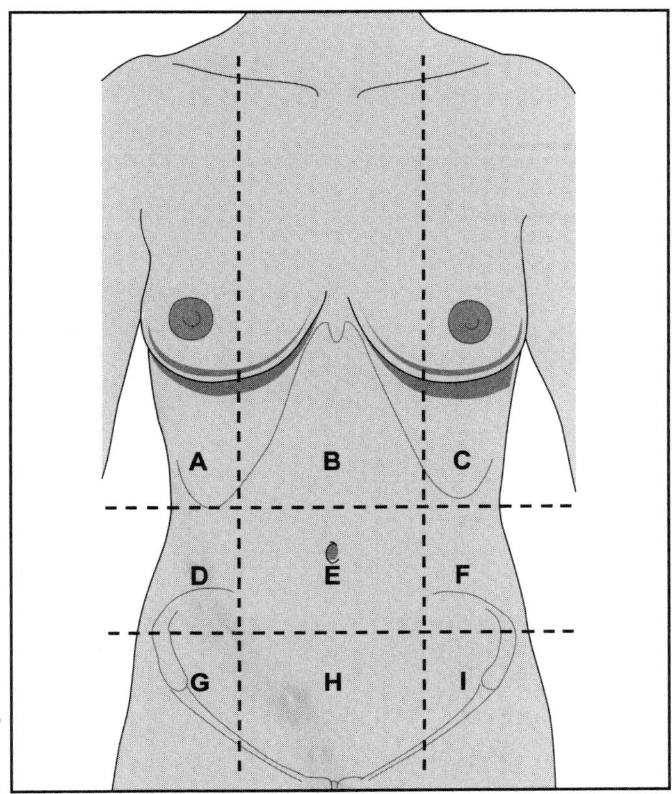

Fig. 1-2

a. Right Hypochondriac

b. Epigastric

c. Left Hypochondriac

d. Right Lumbar

e. Umbilical

f. Left Lumbar

g. Right Iliac

h. Hypogastric

i. Left Iliac

Check Yourself

1. Anatomy is the study of the structures and relationships of the structures of the body. Physiology is the study of the functions of these structures, that is, how they perform their processes. **(Overview of anatomy and physiology)**

2. The subdivisions of anatomy and physiology represent specializations in areas of study.

   a. Systemic anatomy is the study of a specific system of the body such as the digestive or immune system.

   b. Developmental anatomy is the study of the development from a fertilized egg to an adult.

   c. Pathological anatomy is the study of structural changes due to the influence of disease.

   d. Exercise physiology is the study of the functional processes associated with physical exertions.

   e. Endocrinology is the study of the hormonal control of functional processes.

   f. Immunology is the study of defense mechanisms of the body. **(Overview of anatomy and physiology)**

3. The level considered to have the least complexity is the chemical level, followed by the cell, tissue, organ, system and organism level. **(Levels of structural complexity)**

4. The tissues of the human body are epithelial, connective, muscle, and nervous. **(Levels of structural complexity)**

5. An organ is a discrete anatomical structure which contains two or more different types of tissues and performs specific functions. **(Levels of structural complexity)**

6. b **(Levels of structural complexity)**

7. d **(Levels of structural complexity)**

8. a **(Levels of structural complexity)**

9. c **(Levels of structural complexity)**

10. e **(Levels of structural complexity)**

11. Various authors will identify slightly different numbers of common life processes. A minimal list should include: metabolism, responsiveness, movement, growth, and reproduction. Additional processes which are often identified include: boundary maintenance, cellular differentiation, and excretion. **(Life processes)**

12. Homeostasis may be defined as a dynamic equilibrium of the internal conditions of an organism, that is, a state of maintaining certain physiological limits in the organism. Its significance to life is related to the fragile nature of living organisms. Life's processes will only function effectively under relatively narrow

physiological limits. Homeostasis is the maintaince of these limits despite the occurrence of external environmental changes. **(Homeostasis)**

13. As an integral part of the physiological conditions which must be maintained for human life processes, certain requirements exist. They include acceptable levels of nutrients, water, oxygen, temperature, and atmospheric pressure. **(Homeostasis)**

14. Stress is any stimulus (internal or external) which tends to produce an imbalance in the internal environment of the organism. **(Homeostasis)**

15. As stress produces variations in the internal condition of an organism, responses to the resulting imbalances must be made to counteract the changes. Since many different systems must cooperate to maintain the internal environment, communication between theses systems is required. The nervous and endocrine systems are specialized to provide this communication. **(Homeostasis)**

16. Feedback loops include: a control center which determines the setpoint or acceptable range for a particular condition of the internal environment, a receptor which monitors the condition, and an effector which acts to alter the condition. **(Regulation and control mechanisms)**

17. A positive feedback system is one in which the stimulus is intensified or enhanced by the response. One example would be the effects of the hormone oxytocin produced by the hypothalamus during labor contractions. Pressure sensitive receptors in the uterine wall respond to the increasing pressure (stimulus) by activating the release of oxytocin which increases muscular contractions of the uterus, thus increasing the pressure. **(Regulation and control mechanisms)**

18. A negative feedback system is one in which the stimulus is reversed by the response. One example would be the effects of blood pressure on heart rate. As blood pressure rises, pressure sensitive receptors in the walls of the aorta activate the cardiac control centers in the brain. In turn, these control centers respond by sending nerve impulses to the heart which acts to slow the heart rate resulting in a lowered blood pressure. **(Regulation and control mechanisms)**

19. The anatomical position has the subject standing and facing the observer, feet flat on the floor with toes pointing slightly outward and palms of the hands facing forward. **(Anatomical terminology)**

20. The standard anatomical position provides an initial reference to the body. Most directional anatomical terms are in reference to this position so that comprehension of the directional terms requires a knowledge of the anatomical position. **(Anatomical terminology)**

21. c **(Anatomical terminology)**

22. j **(Anatomical terminology)**

23. a **(Anatomical terminology)**

24. f **(Anatomical terminology)**

25. d **(Anatomical terminology)**

26. i **(Anatomical terminology)**

27. b **(Anatomical terminology)**

28. e **(Anatomical terminology)**

29. g **(Anatomical terminology)**

30. h **(Anatomical terminology)**

31. a. frontal

    b. sagital

    c. transverse **(Anatomical terminology)**

32. The principal cavities are the dorsal and ventral cavities. The dorsal is further divided into the cranial and vertebral cavities. The ventral is further divided into the thoracic, abdominal, and pelvic cavities. **(Anatomical terminology)**

33. a. right hypochondriac

    b. epigastric

    c. left hypochondriac

    d. right lumbar

    e. umbilical

    f. left lumbar

    g. right iliac

    h. hypogastric

    i. left iliac **(Anatomical terminology)**

Grade Yourself

Circle the numbers of the questions you missed, then fill in the total incorrect for each topic. If you answered more than three questions incorrectly, you need to focus on that topic. (If a topic has less than three questions and you had at least one wrong, we suggest you study that topic also. Read your textbook, a review book, or ask your teacher for help.)

Subject: The Special Senses

Topic	Question Numbers	Number Incorrect
Overview of anatomy and physiology	1, 2	
Levels of structural complexity	3, 4, 5, 6, 7, 8, 9, 10	
Life processes	11	
Homeostasis	12, 13, 14, 15	
Regulation and control mechanisms	16, 17, 18	
Anatomical terminology	19, 20, 21, 22, 23, 24, 25, 26, 27, 28, 29, 30, 31, 32, 33	

Basic Chemistry of Life

Brief Yourself

Humans, like all living organisms, are composed of chemicals and all of our physiological activity is chemical in nature. For this reason, a familiarity with the language and basic concepts of chemistry is important to the comprehension of anatomy and physiology.

Scientists identify two basic divisions in the composition of the universe: matter and energy. Chemistry might also be termed the study of the interactions of matter. Matter is anything that occupies space and has mass. Mass is the amount of matter contained in an object. We commonly use the term weight interchangeably with mass, but this is not entirely accurate. Weight is the force of gravity that works on mass and gravitational force varies depending upon the distance of the mass from the core of the gravitational field. In other words, while we might weigh 150 pounds while standing on the surface of the earth (relatively near the core of the earth's gravitational field), orbiting in space far from the surface we are essentially "weightless" although our mass is the same in both places.

Although energy is a topic of physics, matter and energy are truly inseparable. Chemical reactions are driven by the energy available in systems in which the reactions occur. Energy is less tangible than matter in that it has no mass and does not occupy space, but it exists as the capacity to do work or to put matter into motion. We measure energy in relation to its effect on matter, with matter as the substance and energy as the mover of the substance. Studies in the science of thermodynamics have taught us that energy is conserved, that is, that energy can neither be created nor destroyed, but is readily converted from one form to another.

Test Yourself

1. What are the two principal kinds of energy? How do they differ?

Questions 2–6 are matching. Match the following form of energy to the examples below:

Possible answers:

2. Ionic currents	a. Chemical energy
3. Visible light	b. Heat energy
4. Boiling water over a burner	c. Radiant energy
5. Extension of a joint by muscle contraction	d. Electrical energy
6. Breaking molecular bonds	e. Mechanical energy

7. What is an atom? What is an element? Explain the relationship between an atom and an element.

8. According to the planetary model of the atom, what are the three sub-atomic particles and where might they be found in an atom?

9. What is the difference between atomic mass of an atom and atomic weight of an element?

10. What is an isotope? What is a radioisotope?

Questions 11–15 are matching. Match the appropriate form in which matter combines to the characteristic.

11. Heterogeneous mixture in which particles tend to settle	a. elemental molecule
12. Two or more atoms of different elements joined by chemical bonds	b. compound molecule
13. Undergoes sol-gel transformation	c. solution
14. Homogeneous mixture in which particles typically do not scatter light nor tend to settle	d. colloid
15. Two or more atoms of the same element joined by chemical bonds	e. suspension

16. What is an electron shell? How are the valence shells related to chemical bonding?

17. How are chemical bonds formed? Distinguish between ionic and covalent bonds and give an example of each.

18. What is a polar molecule and how does it lead to hydrogen (weak electric) bonds?

Questions 19–21 refer to the periodic table.

19. How would you produce a 2.0 molar solution of $MgCl_2$ in water?

20. What is the relationship among the elements found in the first column of the periodic table?

21. What is the name given to the elements in the last column of the periodic table? What is their common characteristic?

22. Since chemical reactions are the result of collisions between molecules, how does energy (heat) and concentration effect chemical reaction rate?

23. What are the three principal types of chemical reactions? Identify each reaction as typically endergonic or exergonic.

24. What are dehydration and hydrolysis reactions? How do they relate to anabolic or catabolic reactions?

25. What is an organic molecule? What three classes of macronutrients are organic?

26. Define an acid, a base, and a salt. Identify a reaction that will demonstrate the relationship between them.

27. Define pH and describe the pH scale.

28. What is the carbonic acid-bicarbonate buffer system in our blood and how does it operate?

29. Water is an inorganic macronutrient. Identify the five essential functions of water in living systems.

30. Define a carbohydrate. Identify the primary classes of carbohydrates. Why are carbohydrates essential nutrients in the body?

31. How do lipids differ from carbohydrates? What is the difference between saturated and unsaturated fats? What are the uses for lipids in the body?

32. Define a protein. Describe the classes of proteins based on their structural organization. Which level of protein structural organization would tend to be the least stable and why?

33. What is a catalyst? What is an enzyme and how might it serve as a biological catalyst?

34. How do deoxyribonucleic acids (DNA) and ribonucleic acids (RNA) differ? What are the functions of each in the body?

35. What is adenosine triphosphate? What is its function in the body?

Periodic Table of the Elements

KEY

Atomic Mass → 12.0111
Symbol → C
Atomic Number → 6
Electron Configuration → $1s^2 2s^2 2p^2$
Selected Oxidation States → -4, +2, +4

Relative atomic masses are based on ^{12}C = 12.00000

New Designation / Former Designation (prior to 1984 IUPAC decision)

Period	1 IA (s-block)	2 IIA (s-block)	3 IIIB	4 IVB	5 VB	6 VIB	7 VIIB	8 VIII	9 VIII	10 VIII	11 IB	12 IIB	13 IIIA	14 IVA	15 VA	16 VIA	17 VIIA	18 0
1	H 1 1.00794																	He 2 4.00260
2	Li 3 6.941	Be 4 9.01218											B 5 10.81	C 6 12.0111	N 7 14.0067	O 8 15.9994	F 9 18.998403	Ne 10 20.179
3	Na 11 22.98977	Mg 12 24.305											Al 13 26.98154	Si 14 28.0855	P 15 30.97376	S 16 32.06	Cl 17 35.453	Ar 18 39.948
4	K 19 39.0983	Ca 20 40.08	Sc 21 44.9559	Ti 22 47.88	V 23 50.9415	Cr 24 51.996	Mn 25 54.9380	Fe 26 55.847	Co 27 58.9332	Ni 28 58.69	Cu 29 63.546	Zn 30 65.39	Ga 31 69.72	Ge 32 72.59	As 33 74.9216	Se 34 78.96	Br 35 79.904	Kr 36 83.80
5	Rb 37 85.4678	Sr 38 87.62	Y 39 88.9059	Zr 40 91.224	Nb 41 92.9064	Mo 42 95.94	Tc 43 (98)	Ru 44 101.07	Rh 45 102.906	Pd 46 106.42	Ag 47 107.868	Cd 48 112.41	In 49 114.82	Sn 50 118.71	Sb 51 121.75	Te 52 127.60	I 53 126.905	Xe 54 131.29
6	Cs 55 132.905	Ba 56 137.33	La-Lu 57–71	Hf 72 178.49	Ta 73 180.948	W 74 183.85	Re 75 186.207	Os 76 190.2	Ir 77 192.22	Pt 78 195.08	Au 79 196.967	Hg 80 200.59	Tl 81 204.383	Pb 82 207.2	Bi 83 208.980	Po 84 (209)	At 85 (210)	Rn 86 (222)
7	Fr 87 (223)	Ra 88 226.025	Ac-Lr 89–103	Unq* 104 (261)	Unp 105 (262)	Unh 106 (263)	Uns 107 (262)	Uno 108	Une 109									

* The systematic names and symbols for elements of atomic numbers greater than 103 will be used until the approval of trivial names by IUPAC.

MASS NUMBERS IN PARENTHESES ARE MASS NUMBERS OF THE MOST STABLE OR COMMON ISOTOPE.

d-block / f-block

Series	d-block								f-block						
Lanthanoid Series	La 57 138.906	Ce 58 140.12	Pr 59 140.908	Nd 60 144.24	Pm 61 (145)	Sm 62 150.36	Eu 63 151.96	Gd 64 157.25	Tb 65 158.925	Dy 66 162.50	Ho 67 164.930	Er 68 167.26	Tm 69 168.934	Yb 70 173.04	Lu 71 174.967
Actinoid Series	Ac 89 227.028	Th 90 232.038	Pa 91 231.036	U 92 238.029	Np 93 237.048	Pu 94 (244)	Am 95 (243)	Cm 96 (247)	Bk 97 (247)	Cf 98 (251)	Es 99 (252)	Fm 100 (257)	Md 101 (258)	No 102 (259)	Lr 103 (260)

Check Yourself

1. Energy is said to be either kinetic or potential. Kinetic energy is expressed as work or motion. For example, a baseball thrown through the air expresses kinetic energy by its motion. Similarly, the constant vibration of tiny particles of matter is kinetic energy. Potential energy is inactive or stored energy due to the position or internal composition of the object. A firecracker has potential chemical energy within, but until the fuse is lit, this energy is not yet expressed. **(Definitions of matter and energy)**

2. d **(Definitions of matter and energy)**

3. c **(Definitions of matter and energy)**

4. b **(Definitions of matter and energy)**

5. e **(Definitions of matter and energy)**

6. a **(Definitions of matter and energy)**

7. An atom is the smallest unit of matter that can enter into a chemical reaction. Atoms are composed of sub-atomic particles in a particular arrangement. Elements are composed of atoms, but the atoms of each element are unique in their composition with regard to their sub-atomic particles. For example, Helium (He) is an element. Only the atoms of He contain 2 positively charged protons, 2 negatively charged electrons and typically 2 neutral neutrons (the number of neutrons may vary slightly). **(Composition of matter)**

8. Although an orbital model of an atom is a more accurate representation of the true condition of an atom, the simplified planetary view is appropriate for instruction in basic chemistry. According to this view, the protons (+) and neutrons (o) are the sub-atomic particles found together in the nucleus of the atom, with electrons (-) orbiting the nucleus at some distance (often called a shell or orbital). **(Composition of matter)**

9. The mass of a sub-atomic particle can be expressed as atomic mass units (amu). Protons and neutrons have virtually the same mass and are defined as having an amu of 1.0. Electrons are approximately 1/2000th the mass of protons or neutrons and as such can be considered to have an effective mass of 0.0 amu. Due to this relatively insignificant mass, they may be ignored in determining the mass of atoms or molecules. Atomic mass for any atom is determined by adding the number of protons and neutrons. A He atom with 2 protons and 2 neutrons would have an atomic mass of 4.0 amu. Atomic weight is a measure of the average of all atoms of an element that reflects the relative proportion of isotopes. For example, most atoms of He have only 2 neutrons, but a few have 3 neutrons. As such, when the atomic masses of all He atoms are averaged, the atomic weight is determined to be 4.00260. **(Composition of matter)**

10. Atoms of the same element must have identical numbers of protons and electrons, but the number of neutrons can vary slightly. Atoms of the same element with different numbers of neutrons are said to be isotopes. Radioisotopes are the more massive isotopes of elements that tend to be less stable and will decay by the emission of alpha or beta particles or gamma rays. **(Composition of matter)**

11. e **(Molecules and mixtures)**

12. b **(Molecules and mixtures)**

13. d **(Molecules and mixtures)**

14. c **(Molecules and mixtures)**

15. a **(Molecules and mixtures)**

16. An electron shell is an area in which electrons may be found within atoms as they orbit around the nucleus. Electron shells exist as energy levels with the energy associated with any electron increasing as it occupies shells that are farther from the nucleus. Since the positive protons attract the negative electrons, in order to orbit farther from the point of attraction, more energy must be utilized. There are 7 identified electron shells, with each having a maximum number of electrons it can hold. The valence shell is the outermost shell that is occupied by an electron for any given atom. It is only the interaction of electrons in the valence shell that produces chemical bonding. **(Chemical bonding)**

17. Chemical bonds are formed by the interactions of the electrons in the valence shells of two or more atoms. The two types of strong chemical bonds are ionic and covalent. Ionic bonds occur as atoms tend to donate or accept one or two electrons in order to produce the more stable configuration of a full outer valence shell, producing atoms with unequal numbers of electrons and protons and thus obtaining an electric charge for the atom. Such a charged atom is called an ion. Ions of different charge will be attracted to one another by the electromagnetic force and will form the ionic bond. NaCl is an example of an ionically bonded molecule. Other atoms with valence shells that will require the loss or addition of several electrons to completely fill their valence shells tend to share electrons. This shuttling of electrons back and forth between the atoms produces a covalent bond. CH_4 is an example of a covalently bonded molecule. **(Chemical bonding)**

18. Polar molecules are those which are formed from covalent bonds in which the atoms do not equally share the electron due to one nuclear center having a larger attractive force than another. For example, in a molecule like water, H_2O, the attraction of the oxygen nucleus with eight protons is much stronger than that of the hydrogen nucleus with but one proton. For this reason, the shared electrons will spend more time orbiting the oxygen than the hydrogen producing weak electromagnetic forces in the molecule. This is common in molecules that have covalent bonds to hydrogen. These weak electromagnetic forces act in a similar fashion to ions, producing weak bonding between polar molecules where the charges are opposite and attractive. This hydrogen bond is very important in determining the nature of many of the complex organic molecules. **(Chemical bonding)**

19. A 1.0 molar solution is defined as 1.0 mole of the solute in 1.0 liter of the solvent. As such, to produce a 2.0 molar solution of $MgCl_2$, we will need to determine the weight of 2 moles of $MgCl_2$ to add to the 1.0 liter of water. A mole of any element is defined as the atomic weight of that element in grams. To determine the weight of a molecule and not an element, it is necessary to determine the molecular weight which is simply the combined atomic weights of each atom of the molecule in the proper proportion. The molecular weight of $MgCl_2$ is determined by taking the atomic weight of Mg (multiplied by 1, since the molecule contains only one Mg) plus the atomic weight of Cl (multiplied by 2, since the molecule contains 2 Cl):

Mg	$24.3050 \times 1 =$	24.3050
Cl	$35.4527 \times 2 =$	70.9054
Mol. Wt.		95.2104

Since we are required to produce a 2.0 molar solution, we will need 95.2104 x 2 = 190.4208 grams. Water is added to this mass as $MgCl_2$ to bring the final volume up to 1.0 liter. **(Molecules and mixtures)**

20. Elements of the same column will contain the same number of electrons in their valence shells and as such will produce similar chemical reactions. The elements of the first column (I) of the periodic table have one electron in their valence shells and as such tend to react by donating this electron to another atom and forming the positively charged ion in an ionic bond. **(Chemical bonding)**

21. The elements in the last (VIII) column of the periodic table are often called the noble gases or inert elements. These elements contain full valence shells and as such are stable and will not produce chemical bonds. **(Chemical bonding)**

22. Heat is a measure of the kinetic energy expressed in the vibration of small particles of matter. The greater the amount of vibration (heat), the greater the likelihood of collisions. In much the same manner, concentration is a measure of the number of particles in a confined space. The greater the number of particles moving in a confined space, the greater the likelihood of collisions. **(Chemical reactions)**

23. The three principal types of chemical reactions are:

Synthesis	A + B	==>	AB
Decomposition	AB	==>	A + B
Exchange	AB + CD	==>	AC + BD

Synthesis increases the complexity of the molecule requiring energy (endergonic). Decomposition produces a decrease in the complexity of the molecule releasing the bound energy (exergonic). The exchange reaction is an energetically neutral reaction as the complexity is unchanged. **(Chemical reactions)**

24. Dehydration and hydrolysis are reactions which utilize a molecule of water in the reaction. Dehydration is a synthesis reaction in which a molecule of water (H + OH) is removed from the substrates to allow a bond to form to produce a product. Similarly, hydrolysis is a decomposition reaction in which a molecule of water is added to the substrate to break a bond producing the products. **(Chemical reactions)**

25. Organic molecules are molecules which contain the element carbon (C). Carbon is an element which forms four covalent bonds. In addition to carbon, the common elements of organic molecules include hydrogen, oxygen, and nitrogen, along with lesser amounts of sulfur and phosphorus. Macronutrients are those that living systems need in large amounts. Organic macromolecules include carbohydrates, lipids, and proteins. **(Organic chemistry)**

26. When acids, bases, and salts are dissolved in water, these chemicals disassociate to produce ions. An acid is a chemical that releases one or more hydrogen ions (H^+) and one or more anions when it disassociates. As such, it is a proton donor. A base is a chemical that releases one or more hydroxide ions (OH^-) and one or more cations. It is a proton acceptor. A salt disassociates into a cation and anion, neither of which are H^+ or OH^-. The following reaction demonstrates their relationship:

HCl +	KOH ==>	KCl +	H_2O	
acid	base	salt	water	**(Inorganic chemistry)**

27. pH is the measure of the acidity or alkalinity of a solution. The pH scale is based on the concentration of H^+ ions in moles per liter. As the concentration of H^+ increases, the concentration of OH^- would decrease and vice versa. A neutral solution, like distilled water, would be at the midpoint of the scale, a pH of 7. This indicates that the solution has 10^{-7} moles of H^+ per liter which is equal to the concentration of OH^- in distilled water. A pH of 6 indicates 10^{-6} of a mole of hydrogen per liter, a tenfold increase in

concentration. So, acids are indicated by lower values on the pH scale and bases indicated by the higher values. **(Inorganic chemistry)**

28. The carbonic acid-bicarbonate buffer system utilizes weak acids and weak bases to automatically compensate for alterations in the pH. If the concentration of H^+ increases in the blood, the following reaction will occur to remove excess H^+:

$$\underset{\text{hydrogen ion}}{H^+} + \underset{\text{bicarbonate ion}}{HCO_3^-} \Longrightarrow \underset{\text{carbonic acid}}{H_2CO_3} \Longrightarrow \underset{\text{water}}{H_2O} + \underset{\text{carbon dioxide}}{CO_2}$$

Similarly, if there is a shortage of H^+ ions, the following reaction will occur to provide additional H^+:

$$\underset{\text{carbonic acid}}{H_2CO_3} \Longrightarrow \underset{\text{hydrogen ion}}{H^+} + \underset{\text{bicarbonate ion}}{HCO_3^-}$$

(Inorganic chemistry)

29. Due to the unique chemical properties of water, it serves five essential functions in living systems: (1) it participates in chemical reactions, specifically dehydration and hydrolysis, (2) has a high capacity for absorption of heat with relatively low changes in temperature; this lessens the impact of environmental fluctuations, (3) has a high heat of vaporization, making it an excellent coolant for the body (perspiration), (4) is a lubricant, and (5) is an excellent solvent due to its polar covalent molecular structure. **(Inorganic chemistry)**

30. Carbohydrates are organic molecules composed of C, H, and O, typically in a 1:2:1 ratio. The primary classes of carbohydrates are monosaccharides and disaccharides, the simple sugars, and polysaccharides, the complex sugars. The principal function of carbohydrates is to produce a chemical energy source for metabolism. A few carbohydrates are used as structural units in humans. One such example is deoxyribose, which is a part of the DNA molecule. **(Organic chemistry)**

31. Although lipids are composed only of C_2, H, and O, they differ from carbohydrates in the ratio of these elements. Lipids have fewer oxygen atoms and thus have fewer polar covalent bonds. The lack of polarity makes them hydrophobic (water fearing) and as such are insoluble in water. A saturated fat is one that has sufficient H atoms so that no double covalent bonds form between its C atoms. Unsaturated fat has insufficient H atoms so that some C atoms will form double bonds with other C atoms. Lipids form many complex molecules with a broad range of functions in the body including protection, insulation, energy storage, structural components (membranes), and hormones and enzyme components. **(Organic chemistry)**

32. Proteins are organic molecules composed primarily of C, H, O, and N. They are formed of chains of amino acids linked together by peptide bonds. They are much more complex in structure than carbohydrates or lipids and have a wide range of functions in the body. Due largely to the polar covalent molecules prevalent in proteins, hydrogen bonding produces additional levels of structural organization of these complex molecules. The primary level of structural organization is given by the order of amino acids in the polypeptide chain. The secondary level of structure commonly includes the alpha helix or beta pleated sheet. The tertiary level of structure occurs as additional hydrogen bonding produces a folding of the helix or sheet. The quartenary level of structure occurs as two or more polypeptides associate together. The greater the complexity of structure, the more unstable the macromolecule becomes making the quartenary and tertiary levels of structure the most likely to be denatured. **(Organic chemistry)**

33. A catalyst is a chemical which can serve to lower the activation energy required for a reaction and properly orient substrates for the production of a product. Catalysts are unchanged in the chemical reaction that they support, and thus are reusable. Enzymes are biological catalysts. Most of our biological catalysts consist of a complex protein part (apoenzyme) and a nonprotein part (cofactor). Most of our metabolism is controlled by our enzymes. Enzymes are thought to function by producing attracting "active sites" on their surfaces which bring the substrates into close proximity and forming a temporary enzyme-substrate complex. This enhances the development of products which are rapidly released allowing the enzyme to be reused. **(Organic chemistry)**

34. DNA has the sugar deoxyribose as part of its sugar-phosphate chain while RNA has the sugar ribose. DNA forms a double-helix while RNA is single-stranded in humans. DNA contains four nitrogenous bases, two single-ring bases called purines (adenine and guanine) and two double-ring bases called pyrimidines (thymine and cytosine). RNA does not contain thymine, but does utilize another pyrimidine in its place, uracil. Nucleic acids are huge molecules that serve as information molecules in living systems. DNA houses our genes, or genetic information that determines which traits we inherit by controlling the protein synthesis and thus controlling metabolism. RNA acts as a messenger and enzyme in relaying the instructions on the DNA to the cell metabolic machinery. **(Organic chemistry)**

35. ATP is often called the energy molecule. It is the only molecule with which our cells can do work. It is composed of an adenine base attached to a ribose sugar with three phosphate groups attached, the last two phosphate groups containing high energy chemical bonds. Cellular respiration is directed toward the production of ATP which we utilize to power our metabolic work. **(Organic chemistry)**

Grade Yourself

Circle the numbers of the questions you missed, then fill in the total incorrect for each topic. If you answered more than three questions incorrectly, you need to focus on that topic. (If a topic has less than three questions and you had at least one wrong, we suggest you study that topic also. Read your textbook, a review book, or ask your teacher for help.)

Subject: Basic Chemistry of Life

Topic	Question Numbers	Number Incorrect
Definitions of matter and energy	1, 2, 3, 4, 5, 6	
Composition of matter	7, 8, 9, 10	
Molecules and mixtures	11, 12, 13, 14, 15, 19	
Chemical bonding	16, 17, 18, 20, 21	
Chemical reactions	22, 23, 24	
Organic chemistry	25, 30, 31, 32, 33, 34, 35	
Inorganic chemistry	26, 27, 28, 29	

Cells—Structure and Function

3

Brief Yourself

Cells are the fundamental units of living things, whether they are the generalized single-celled organism or the more specialized components of the complex multicellular organisms. The cell is the minimum amount and organization of matter that has the characteristics of life. Cells must perform all of the physiological processes necessary to maintain homeostasis in their environment in a rapidly changeable world.

Chemically, cells are composed primarily of carbon, oxygen, hydrogen, nitrogen, and trace amounts of several other elements. It is in the complex arrangement of these elements that the special characteristics of life lie. Cells are dramatically diverse in their size, shape, and specialization (especially in multicellular organisms). Human cells range from a typical size of about 10 μm in diameter to 100 μm in a fertilized egg. Some of our skeletal muscle cells may be as much as 30-40 cm in length and lengths of well over a meter seen in certain nerve cells. Many possibilities for cell-shapes exist, such as spherical, flat and wafer-like, cuboidal, or even stellate.

Despite the fact that no two cells are exactly alike, cells do have many common components and functional features. For this reason, we typically examine a model of a generalized or composite cell to identify these characteristics.

Test Yourself

1. Identify the four concepts of the cell theory.

2. Define a cell and identify its three principal parts.

3. What is a phospholipid and what is its importance to the structure of the plasma membrane of a cell?

4. What is the function of glycolipids and glycoproteins of the plasma membrane? What is the function of cholesterol?

5. What are the two types of plasma membrane proteins and the functions associated with them?

6. What is selective permeability of plasma membranes and how is it controlled?

7. What is the fundamental difference between passive movement and active movement across plasma membranes?

Questions 8–17 are matching. Match the following with the appropriate definition:

8. The net diffusion of water through a selectively permeable membrane
9. Engulfing small particles in droplets of extracellular fluid, bringing them into the intracellular fluid in developed vesicles
10. Particles forced across membranes due to differences in pressure on either side of the membrane
11. Discharge of particles from the intracellular fluid to the extracellular fluid
12. Plasma membrane proteins that bind to ions and then denature to produce a forced movement of these particles across the membrane, ATP is required to produce the protein denaturing
13. Engulfing large particles by means of plasma membrane alterations (pseudopodia) to move particles into the intracellular fluid within developed vesicles
14. An even distribution of particles in which continued exchange produces no net change in concentration
15. Movement of particles through specialized carrier membrane proteins in which no cellular energy is required
16. Random movement of particles due to their kinetic energy that will eventually produce an even distribution of the particles in the available space
17. The difference in the distribution of particles within the available space, such that the concentration of particles will be lower in one area and higher in another

a. Simple diffusion
b. Concentration gradient
c. Equilibrium
d. Osmosis
e. Filtration
f. Facilitated diffusion
g. Phagocytosis
h. Pinocytosis
i. Solute pumps (ion pumps)
j. Exocytosis

18. Identify the effects of an isotonic, hypertonic, and hypotonic solution on red blood cells. Why doesn't tonicity effect most other cells in the same fashion as the RBC?

19. Define cytosol. What is its chemical composition?

20. What are ribosomes and where may they be found in cells? What is their general function?

21. What is endoplasmic reticulum? What two types of ER exist and what are their functions?

22. What is the Golgi apparatus? What is its function?

23. How do lysosomes and peroxisomes differ in structure and function?

24. Why are mitochondria called the "power-house" of the cell?

25. How do microfilaments and microtubules differ in structure and function?

26. Describe the structure and function of the nucleus of the cell. What is a chromosome?

27. The first stage of protein synthesis is transcription. Describe the events of transcription from its inception to its completion. Where does it occur and what is its product?

28. Translation occurs in three phases. Identify these phases and the events of each.

Questions 29–37 are matching questions. Match the following phases of the life cycle of a cell with its characteristics.

29. Stage of mitosis in which karyokinesis (the division of chromosomes) occurs	a. Interphase
30. A preparation for mitosis in which the necessary enzymes and proteins are synthesized	b. Growth 1 sub-phase
31. The continuous process of cellular division	c. Synthetic sub-phase
32. Stage of mitosis in which the chromosomes are aligned along the equator of the cell	d. Growth 2 sub-phase
33. The total period from cell formation to the beginning of cell division	e. Mitosis
34. Stage of mitosis in which cytokinesis occurs	f. Prophase
35. Rapid growth of a cell including its maturation and specialization	g. Metaphase
36. Stage of mitosis in which chromosomes condense and are attached to the developing spindle apparatus	h. Anaphase
37. DNA is duplicated in preparation for mitosis	i. Telophase

Check Yourself

1. Cell theory has been developing since the 1800s. Today's understanding of the cell theory includes the following four concepts:

    a. A cell is the basic structural and functional unit of a living organism.

    b. An organism may be defined and understood by the comprehension of the individual and collective activities of its cells.

    c. Cell components and chemistry are complementary in that the biochemical activities of cells are determined by the specific common subcellular structures of cells.

    d. The continuity of life is cellular in nature. **(Overview of cellular basis of life)**

2. A cell is the basic structural and functional unit of life. The three major parts of the generalized human cell are:

    a. the plasma membrane.

    b. the cytoplasm which contains the organelles and inclusions.

    c. the nucleus which contains the nucleolus and nucleoplasm. **(Overview of cellular basis of life)**

3. Phospholipids are complex lipids that are similar to triglycerides, but with the loss of one fatty acid chain and the addition of a phosphate group. This produces a molecule with a polar "head" (the phosphate group) and a nonpolar "tail" (the fatty acid chains). Between the extracellular and intracellular fluid compartments, these molecules orient to produce a bilayer as the polar heads move toward water and the nonpolar tails move away from the water. **(Plasma membrane structure)**

4. Glycolipids and glycoproteins are complex molecules that include polysaccharide chains that extend into the extracellular fluid. This glycocalyx is important for adhesion and communication among cells. Cholesterol serves to stabilize the phospholipid bilayer. **(Plasma membrane structure)**

5. Plasma proteins are either peripheral or integral, with peripheral proteins attached to the inner or outer surface of the membrane and integral proteins embedded in the membrane and extending to communicate with both the extracellular and intracellular fluid. Plasma proteins serve as ion channels (pores), carriers (pumps), receptors, enzymes, anchors to the cytoskeleton, and identification markers. **(Plasma membrane structure)**

6. Selective permeability is the regulation of the movement of particles across the plasma membrane. Permeability is controlled passively by the chemical characteristics of the membrane and material, as well as, actively by the use of the cell's energy to overcome the chemical characteristics. The chemical characteristics important to membrane permeability are:

    a. Lipid solubility—Substances that dissolve in lipids pass easily through the phospholipid membrane.

    b. Size—Large particles can not pass through the plasma membrane.

c. Charge—Only very small uncharged particles can pass through the plasma membrane. Charged particles of small size must move through special pores produced by channel proteins.

d. Carriers—Specialized integral proteins act as carriers or pumps to move some small particles, both charged and uncharged, and produces a permeability for these particles. **(Plasma membrane structure)**

7. The fundamental difference is the requirement of cellular energy in the processes. In passive movement, physical mechanisms will produce the movement of certain particles across the membrane with no cellular energy required. Active movement requires the utilization of energy, produced by splitting ATP in order to produce the movement of certain particles across the membrane. **(Movement of particles across the plasma membrane)**

8. d **(Movement of particles across the plasma membrane)**

9. h **(Movement of particles across the plasma membrane)**

10. e **(Movement of particles across the plasma membrane)**

11. j **(Movement of particles across the plasma membrane)**

12. i **(Movement of particles across the plasma membrane)**

13. g **(Movement of particles across the plasma membrane)**

14. c **(Movement of particles across the plasma membrane)**

15. f **(Movement of particles across the plasma membrane)**

16. a **(Movement of particles across the plasma membrane)**

17. b **(Movement of particles across the plasma membrane)**

18. A normal saline solution (0.9% NaCl) is isotonic with respect to the intracellular fluid of the RBC, so no net movement of water would occur and the cell would appear unchanged. In a hypertonic solution which has a higher percentage of solute than normal saline, water would diffuse from the RBC into the extracellular fluid causing the cell to shrivel (crenation). In a hypotonic solution with a lower percentage of solute than normal saline, water would diffuse into the RBC from the extracellular fluid causing the cell to swell and eventually burst (hemolysis). Most cells can control the entry or exit of water through their water pores, but RBC's are anucleate cells and as such can not control the plasma membrane channel proteins. **(Movement of particles across the plasma membrane)**

19. Cytosol is the intracellular fluid in which organelles and inclusions are suspended and solutes are dissolved. It is composed primarily of water (70-90%) with inorganic and small organic substances dissolved and larger particles suspended. **(Cytosol)**

20. Ribosomes are small granules that contain ribosomal RNA and associated proteins. They may be found floating freely in the cytoplasm or attached to endoplasmic reticulum. They function in the synthesis of proteins. **(Organelles)**

21. Endoplasmic reticulum is a network of membrane enclosed channels (cisternae) which appears to be continuous with the nuclear membrane and may connect to the plasma membrane. Rough ER is studded with ribosomes and is a site for storage of newly synthesized proteins and may be involved with additional synthesis in the production of complex proteins (glycoproteins or lipoproteins). The smooth ER is a site of synthesis of complex lipids (phospholipds or steroids). **(Organelles)**

22. Golgi apparatus is an organelle that consists of flattened sacs (producing cisternae) typically found near the nucleus of the cell. It is involved with the production of lysosomes and secretory vesicles in which proteins and lipids that have been received from the ER are sorted, processed, and packaged. **(Organelles)**

23. Lysosomes are vesicles containing powerful digestive enzymes which serve to digest particles brought in by phagocytosis and also act to dissolve and recycle damaged cell structures. Peroxisomes are similar in structure but contain enzymes such as catalase that act to oxidize various cellular toxins. **(Organelles)**

24. Mitochondria are organelles specialized for aerobic respiration and as such provide the majority of ATP in human cells. **(Organelles)**

25. Microfilaments are thin strands of contractile proteins, such as actin, that typically attach to the proteins of the plasma membrane and are involved in cell motility and shape. Microtubules are long, flexible tubes formed of globular proteins and are considered to be controlled by the centrosomes. In addition to aiding in the shape and support of cells, they form tracks for the movement of substances and organelles through the cytosol. **(Organelles)**

26. The nucleus is the largest organelle of the cell and contains the hereditary information in the form of the genes on the DNA. It is surrounded by the nuclear envelope, a very porous membrane. Nucleoli are spherical bodies found in the nucleus where ribosomes are being produced. A chromosome is a long DNA molecule coiled and packed into a compact structure with associated proteins called histones. **(Nucleus)**

27. Transcription is the process of copying a portion of the DNA, a gene that codes for a particular messenger RNA and as such, codes for a particular protein. The process occurs in the nucleus of the cell and begins with an enzyme, RNA polymerase, identifying the beginning of the gene and opening the double helix for copying. One side of the DNA, the sense strand, is used as a template for the production of a single-stranded mRNA molecule. The triplets of base sequences on the DNA code for the sequence of amino acids to be used in the production of the protein to be synthesized. These triplets are copied as a mirror image called codons on the developing mRNA strand. When the gene is copied, the mRNA strand is released and the DNA strands reassociate. The introns of the mRNA are removed by mRNA splicing and the mRNA diffuses out of the nucleus into the cytoplasm. **(Protein Synthesis)**

28. Translation begins with initiation, the binding of a small ribosomal subunit to the start codon of the mRNA followed by the binding of the large ribosomal subunit. This exposes the codon of the mRNA for binding with the anti-codon of tRNA. Elongation continues with tRNA binding to a particular amino acid and bringing it to the ribosome. The anti-codon of the tRNA will recognize and bind to the complementary codon on the mRNA. Once a tRNA is bound, the ribosome will move exactly three bases (one codon) along the mRNA. This allows the bonding of the second tRNA, (with the appropriate anti-codon) carrying its amino acid. The contiguous amino acids will form peptide bonds. As the peptide bonds form, the tRNA's are released to be reused, the ribosome moves along the mRNA and the peptide chain grows producing a protein. Termination occurs when the ribosome reaches a "stop" codon at which time the small and large ribosomal subunits will disassociate. **(Protein synthesis)**

29. h **(Cell life cycle and division)**

30. d **(Cell life cycle and division)**

31. e **(Cell life cycle and division)**

32. g **(Cell life cycle and division)**

33. a **(Cell life cycle and division)**

34. i **(Cell life cycle and division)**

35. b **(Cell lifecycle and division)**

36. f **(Cell life cycle and division)**

37. c **(Cell life cycle and division)**

Grade Yourself

Circle the numbers of the questions you missed, then fill in the total incorrect for each topic. If you answered more than three questions incorrectly, you need to focus on that topic. (If a topic has less than three questions and you had at least one wrong, we suggest you study that topic also. Read your textbook, a review book, or ask your teacher for help.)

Subject: Cells—Structure and Function

Topic	Question Numbers	Number Incorrect
Overview of cellular basis of life	1, 2	
Plasma membrane structure	3, 4, 5, 6	
Movement of particles across the plasma membrane	7, 8, 9, 10, 11, 12, 13, 14, 15, 16, 17, 18	
Cytosol	19	
Organelles	20, 21, 22, 23, 24, 25	
Nucleus	26	
Protein synthesis	27, 28	
Cell life cycle and division	29, 30, 31, 32, 33, 34, 35, 36, 37	

Tissues and Membranes

4

Brief Yourself

In living systems, multicellular organisms have certain advantages. Single-celled organisms must be relative generalists as the individual cell must perform all of the physiological processes necessary to sustain life equally well. Multicellular organisms have cells which will form tight colonies that function cooperatively, allowing specialization in functions of certain cells which increases the efficiency of performing life processes. There are also hazards due to this specialization of function to the multicellular organism. The loss of any one community of cells that produces an indispensable function can destroy the total organism.

A tissue is defined as a group of closely associated cells of similar structure that perform a common function. In the human body, we identify four primary tissues; epithelial, connective, muscle, and nervous. Each has numerous subclasses or varieties, which are described in histology, the study of tissues. In addition, the structure and properties of the extracellular material that surrounds the tissue cells and the connections between cells helps to define tissue structure and function.

Test Yourself

1. Although glycocalyx helps cells adhere to one another, special plasma membrane connections exist between cells. Identify the three basic types and state how they differ in structure and function.

2. Epithelial tissues is often called the "covering" tissue. What are the five basic functions of epithelium?

3. What are the two principle types of epithelium?

4. Identify and describe the four basic classes of epithelium based upon cell shape.

5. Identify and describe the three basic classes of epithelium based upon the number of layers.

Questions 6–13 are matching questions. Match the following epithelial tissue with an example of its location and function.

6. Lining most of the digestive tract; absorption and secretion — a. Simple squamous
7. Lining ducts of larger sweat glands, mammary glands; protection — b. Simple cuboidal
8. Lining of trachea and upper respiratory tract; secretion of mucus — c. Simple columnar
9. Lining of mouth, esophagus, epidermis of skin; protection — d. Pseudostratified
10. Rare, found in male urethra; protection and secretion — e. Stratified squamous
11. Alveoli of lungs, capillaries; allows diffusion and filtration — f. Stratified cuboidal
12. Lining ureters and urinary bladder; permits distention — g. Stratified columnar
13. Lining kidney tubules and small glands; secretion and absorption — h. Transitional

Questions 14–21 are matching questions. Match the following with its definition:

14. Simple squamous epithelium that lines the lumen of the heart, blood vessels, and lymphatic vessels — a. Goblet cell
15. Membrane specialization of epithelium consisting of macrotubules — b. Secretion
16. Tough, waterproof protein that is very resistance to damage by friction — c. Absorption
17. Simple squamous epithelium that forms serous membranes — d. Keratin
18. Small, finger-like extensions of the apical cell membrane — e. Endothelium
19. Intake of fluids or particles through epithelial membranes — f. Mesothelium
20. Specialized columnar epithelial cell for the production and secretion of mucus — g. Microvilli
21. Production and release of a fluid from a gland cell — h. Cilia

22. Identify and describe the general structural characteristics of epithelial cells.

23. What is a gland? How do exocrine and endocrine glands differ?

24. How do the general features of connective tissues differ from epithelial tissues?

25. What are the two principal types of connective tissue cells and what specific cells do they include?

Questions 26–36 are matching. Match the following classes of connective tissue with their location and functions.

26. Found in tendons, ligaments; parallel collagen fibers resist force unidirectionally	a. Areolar
27. Found distributed beneath epithelia, surrounds capillaries; wraps and cushions organs, holds and conveys tissue fluid	b. Adipose
28. Found primarily within the blood vessels; transports gases, nutrients, wastes and forms components of the immune system	c. Reticular
29. Found in the dermis of the skin and fibrous capsules; random collagen fibers resist force in many directions	d. Dense regular
30. Found in the walls of the aorta and bronchi; provides strength with stretch and resilience	e. Dense irregular
31. Found in the inter-vertebral discs, pubic symphysis, and knee; provides tensile strength and shock absorption.	f. Elastic
32. Found in the bones; provides support, protection, levers for movement, mineral storage and bone cell formation	g. Hyaline cartilage
33. Found under the skin and around visceral organs; stores lipids, insulates, shock absorption and protection	h. Elastic cartilage
34. Found in the external ear; maintains shape of structure while providing flexibility	i. Fibrocartilage
35. Found in the lymphoid organs; fibers form a soft internal skeleton and support other cell types	j. Osseous
36. Found in the embryonic skeleton, ends of bones, costal cartilage; supports and reinforces, protects and cushions	k. Blood

37. Describe the structure of an osteon of compact bone.

38. Describe and differentiate between serous, mucous, and synovial membranes.

39. What are the classifications of muscle tissue? How do the cells of these tissues differ?

40. Identify the two principal types of nerve cells and describe their function.

41. Identify the steps in the repair of a nonextensive skin wound.

Check Yourself

1. The plasma membrane connections between cells include the following:

   a. Tight junctions: adjacent membranes fuse to produce a "water-tight" connection between cells that prevents the flow of extracellular fluid in this area.

   b. Spot desmosomes: act as mechanical junctions between cell membranes with extended glycoproteins that produce anchors between the membranes. They are particularly strong connectors.

   c. Gap junctions: hollow cylinders of protein channels extending between the membranes of adjacent cells that allows for direct passage of materials between cells. **(Cell junctions)**

2. Epithelium is specialized for the following functions:

   a. Protection

   b. Absorption

   c. Secretion

   d. Filtration

   e. Excretion **(Epithelium)**

3. Epithelium is classified as covering or lining epithelium and glandular epithelium. **(Epithelium)**

4. Epithelial cells are classified by shape as:

   a. Squamous: flattened, thin and tile-like

   b. Cuboidal: thick, hexagonal or cube-shaped

   c. Columnar: tall, cylindrical

   d. Transitional: readily vary shape from flat to columnar, often have a "domed" apical membrane. **(Epithelium)**

5. Epithelial cells are also classified by layering as:

   a. Simple: a single layer, where single cells have both apical and basal faces.

   b. Stratified: multiple layers, where no single cell has both apical and basal faces.

   c. Pseudostratified: all cells have a basal face with some cells having apical faces but other cells being shorter and failing to reach the apical surface. This provides a false impression that multiple layers exist. **(Epithelium)**

6. c **(Epithelium)**

7. f **(Epithelium)**

8. d **(Epithelium)**

9. e **(Epithelium)**

10. g **(Epithelium)**

11. a **(Epithelium)**

12. h **(Epithelium)**

13. b **(Epithelium)**

14. e **(Epithelium)**

15. h **(Epithelium)**

16. d **(Epithelium)**

17. f **(Epithelium)**

18. g **(Epithelium)**

19. c **(Epithelium)**

20. a **(Epithelium)**

21. b **(Epithelium)**

22. Epithelium has a particularly dense cellularity with little space for extracellular material (matrix) between its cells and has numerous membrane junctions. Epithelial membranes have an apical (free) face exposed to the external surface of the body or to an internal cavity (lumen) as well as a basal (fixed) face attached to the basement membrane. The apical face often has membrane specializations such as cilia or microvilli. The dense cellularity of epithelial cells, especially in stratified tissue, produces avascularity. The basal face of epithelium produces a basal lamina, a matrix of glycoproteins that help to adhere the tissue to the cells below. Epithelium is a highly regenerative tissue. **(Epithelium)**

23. Glands consist of one or more specialized epithelial cells that secrete substances into ducts, onto a surface, or into the blood. Endocrine glands secrete directly into the extracellular fluid and typically its secretions enter the blood stream. Exocrine glands secrete onto surfaces, usually by way of a tube-like duct. **(Glands)**

24. Connective tissues consist of cells, ground substance, and fibers. Unlike epithelial cells, there is typically a good deal of space between the cells of connective tissue and these cells produced a functionally important matrix. It typically does not form a lining, and with the exception of cartilage, is a highly vascularized tissue. **(Connective tissues)**

25. Connective tissue cells are said to be either fixed or mobile (wandering). Fixed cells are typically the producers of the matrix that help provide the functional characteristics of the tissue. Fixed cells include the fibroblasts, chondrocytes, hematocytoblasts, and osteoblasts. Mobile cells include the erythrocytes, leukocytes, mast cells, macrophages, and plasma cells. **(Connective tissues)**

26. d **(Connective tissues)**

27. a **(Connective tissues)**

28. k **(Connective tissues)**

29. e **(Connective tissues)**

30. f **(Connective tissues)**

31. i **(Connective tissues)**

32. j **(Connective tissues)**

33. b **(Connective tissues)**

34. h **(Connective tissues)**

35. c **(Connective tissues)**

36. g **(Connective tissues)**

37. An osteon is a cylinder of compact (cortical) bone growth. Surrounding a central blood vessel, concentric rings of osteocytes exists in small cavities called lacuane. These cells create the bony matrix in layers called lamellae. The osteocytes are stellate in shape with extensions providing connections for membrane junctions between the cells. This allows for the passage of nutrients and wastes through the layers of cells to support those farther from the central blood vessel. The small openings in the matrix to support these extensions are called canaliculi. **(Connective tissues)**

38. Serous membranes line body cavities that do not open directly to the exterior. They are thin layers of areolar connective tissue covered by mesothelium. The layer attached to the cavity wall is called parietal and the layer attached to the organs inside the cavity is called visceral. Mucous membranes line the body cavities that open directly to the exterior and contain mucus producing cells such as goblet cells. Synovial membranes line the cavity of freely movable joints and secrete synovial fluid. They contain no epithelium and are composed primarily of areolar and adipose connective tissues. **(Membranes)**

39. Muscle is classified as skeletal (striate), smooth, and cardiac. Skeletal muscle cells are greatly elongated, multinucleate cells with highly specialized myofibrils organized into sarcomeres, which produces a striped appearance. Cardiac muscle cells are uninucleate and often branched. They are not as elongated as striate cells, but do show a characteristic striped pattern. They are coupled end to end with adjacent cells by means of a specialized membrane junction similar to gap junctions and spot desmosomes, called an intercalated disc. Smooth muscle cells are uninucleate and nonstriated. These cells are spindle-shaped, thickest in the center where the nucleus is located and tapering at both ends. Like cardiac cells, they are sometimes connected to adjacent cells by means of gap junctions. **(Muscle)**

40. Nerve cells are classified as neurons or neuroglials. Neurons are specialized for propagating impulses and typically consist of a soma with specialized extensions called dendrites and an elongated process called an axon. Neuroglial cells do not generate or conduct impulses. Instead, they function in the support of neurons. **(Nervous)**

41. When severed blood vessels bleed, inflammatory chemicals are released which tend to dilate the capillary networks in the area, producing greater permeability to allow leukocytes, fluid, clotting proteins, and other plasma proteins into the area. Clotting occurs and granulation tissue is formed. Capillary nets will invade the clot, reintroducing blood flow. Fibroblasts invade the region producing fibers to help anchor and bridge the wound. Macrophages provide a "clean up" by phagocytizing dead cells and debris. Surface epithelial cells multipy to cover the granulation tissue and the fibrosed scar contracts. **(Tissue repair)**

Grade Yourself

Circle the numbers of the questions you missed, then fill in the total incorrect for each topic. If you answered more than three questions incorrectly, you need to focus on that topic. (If a topic has less than three questions and you had at least one wrong, we suggest you study that topic also. Read your textbook, a review book, or ask your teacher for help.)

Subject: Tissues and Membranes

Topic	Question Numbers	Number Incorrect
Cell junctions	1	
Epithelium	2, 3, 4, 5, 6, 7, 8, 9, 10, 11, 12, 13, 14, 15, 16, 17, 18, 19, 20, 21, 22	
Glands	23	
Connective tissues	24, 25, 26, 27, 28, 29, 30, 31, 32, 33, 34, 35, 36, 37	
Membranes	38	
Muscle	39	
Nervous	40	
Tissue repair	41	

Integumentary System

Brief Yourself

A group of two or more types of tissues working in concert to perform a specific function is an organ. Of all the organs of the human body, the skin is the most exposed to infection and injury from its interactions with the external environment. Serving its primary function of protection, it is the outer boundary to the human organism. Waterproof, stretchable, and capable of repairing itself, the skin is the largest organ of the body.

A group of organs working together to perform specific functions is a system. The skin is the largest organ of the integumentary system, but this system also includes associated organs and derivatives of the skin such as hair, nails, glands, and specialized nerve endings. It has been estimated that a single square centimeter (cm^2) of the integument contains as much as 70 cm of blood vessels, over 100 glands, and well over 200 sensory receptors.

Test Yourself

1. What are the two principal layers of tissue that compose the skin? What layer of integumentary tissue is found just below the skin?

Questions 2–5 are matching. Match the following epidermal cells with their characteristics.

2. Interacts with leukocyte helper-T cell in immune responses
3. Produces protein and serves primarily in protection; makes up 90% of epidermis
4. Functions in sensory reception, found in the deepest layer of epidermis
5. Produces a brown-black pigment that absorbs UV light

a. Keratinocytes
b. Melanocytes
c. Merkel cell
d. Langerhans cell

6. Identify the four layers of the normal epidermis from profound to superficial.
7. The glabrous skin of the palms of the hands and soles of the feet has an additional layer of epidermis. Identify this layer and explain why it is needed.

8. What are the two principal layers of the dermis and how do they differ?

9. What is the function of dermal papillae? What is an epidermal ridge?

10. What factors produce skin color? What is an albino?

Questions 11–14 are matching. Match the following sensory receptor of the skin with its location and function.

11. Found in the reticular layer; responsible for deep pressure sensations	a. Meissner's corpuscle
12. Found in the reticular layer; responsible for sensation associated with movement of hairs	b. Pacinian corpuscle
13. Found in the papillary layer; responsible for light touch sensations	c. Free nerve endings
14. Found primarily in the papillary layer; responsible for sensations associated with pain	d. Root hair plexus

15. Identify and describe the types of sudoriferous glands found in the integument. Differentiate between them on the basis of function.

16. What are sebaceous glands? Where are they found and what is their function?

17. Describe the principal parts of a nail. What is its function?

Questions 18–25 are matching. Match the following anatomical structures of a hair with its description.

18. Outermost layer of a hair; single layer of thin, highly keratinized cells	a. Shaft
19. The superficial portion of the hair which projects from the surface of the skin	b. Root
20. Enlarged, layered structure serving as the anchor of the hair	c. Medulla
21. The major part of the hair containing elongated, pigmented cells	d. Cortex
22. The deep portion of the hair that extends into the dermis	e. Cuticle
23. An epithelial structure from which the hair develops	f. Follicle
24. The wall of the hair follicle consisting of connective tissue and epithelial tissue layers	g. Root sheath
25. Inner rows of polyhedral cells containing pigment and air spaces	h. Bulb

26. Describe the physiological processes of the skin which serve in thermoregulation.

27. In its protective function, the skin produces three barriers. Identify and describe these barriers.

28. Why is the skin often considered to be a layer of cutaneous sensory reception?

29. What is the excretory function of the integument? What is its synthesis function?

30. Describe the hypodermis and its functions.

31. Describe the stages involved in the healing of a deep skin wound.

Questions 32–39 are matching. Match the following term with its definition.

32. Yellowed appearance in the whites of the eye and skin due to a buildup of bilirubin
33. The absence of melanocytes from a patch of skin
34. Accumulation of melanocytes in one area
35. Increased production of a melanocyte stimulating hormone producing an abnormal increase in skin coloration
36. Natural age-related atrophy of hair follicles resulting in loss of hair as its rate of shedding is greater than its rate of replacement
37. Bluish tint to the skin due to a lack of oxygen in the hemoglobin causing the loss of its typical red color
38. Orientation of collagen fibers in one predominant direction in the dermis
39. Excessive production of androgens resulting in abnormally thick hair production on the lip, chin, chest, thighs, and abdomen

a. Cleavage line
b. Freckle
c. Hirsutism
d. Alopecia
e. Vitiligo
f. Jaundice
g. Cyanotic
h. Addison's disease

Check Yourself

1. The skin is composed of an outer layer, the epidermis and a deeper layer, the dermis. The skin rests upon the hypodermis, a subcutaneous tissue layer. **(Skin)**

2. d **(Epidermis)**

3. a **(Epidermis)**

4. c **(Epidermis)**

5. b **(Epidermis)**

6. From the most profound to the most superficial, the strata of the epidermis are: stratum basale (germinativum), stratum spinosum, stratum granulosum, and stratum corneum. **(Epidermis)**

7. The stratum lucidum is found primarily in the palms of the hands and the soles of the feet. It adds additional thickness to the skin where greater friction typically occurs. **(Epidermis)**

8. The dermis consists of two principal portions. The more superficial portion consists primarily of areolar connective tissue with many fine elastic fibers. The surface area between the more superficial epidermis and the dermis is greatly increased by finger-like extensions of the dermis called dermal papillae which gives rise to the name of this portion, the papillary region. The deeper portion of the dermis is the reticular region which consists of dense, irregular connective tissue intertwined with coarse elastic fibers. This is the region from which the hair and glands of the skin arise. **(Dermis)**

9. Dermal papillae serve to dramatically increase the surface area between the epidermis and the dermis. The epidermis is composed primarily of a specialized stratified squamous epithelium which, like all such tissue, is avascular. Most papilla contains a capillary loop providing increased exchange with the epidermis. On the ventral surfaces of the hands and feet, the papillae are arranged in specific and genetically determined patterns called epidermal ridges. These function to increase the surface friction to enhance the gripping ability of the hands and feet. Imprints of these patterns from the fingers are fingerprints. **(Dermis)**

10. Skin color is produced by three pigments; melanin, carotene, and hemoglobin. Melanin is produced in the melanocytes and dispersed to the keratinocytes. The difference in skin color due to melanin is typically not due to a difference in the number of melanocytes, but in the amount of pigment that they produce. Increased exposure to UV light tends to increase their activity and darken the skin. Carotene is a precursor of vitamin A and tends to accumulate in the cells of the stratum corneum. The amount of dilation of the capillary nets in the dermal papillae adds to the hue of the skin by exposing greater or lesser blood flow in the immediate area of the skin. An albino is a person unable to synthesize melanin. **(Dermis)**

11. b **(Dermis)**

12. d **(Dermis)**

13. a **(Dermis)**

14. c **(Dermis)**

15. The principal types of sudoriferous glands are eccrine, apocrine, ceruminous, and mammary. Eccrine are the most common and function primarily in thermoregulation by the production of sweat. Apocrine are confined to the anogenital and axillary regions of the body and produce sweat that contains fatty substances and proteins and may be analogous to sexual scent glands of animals. Ceruminous are modified apocrine glands found in the lining of the external acoustic meatus that produce cerumen (ear wax). Mammary are specialized sweat glands found in females that secrete milk. **(Glands)**

16. Sebaceous glands are oil glands and produce a lipid based secretion called sebum. Found in the dermis, these holocrine type glands typically have ducts that lead to a hair follicle so that the sebum migrates to the surface along the hair. The function of sebum is to serve as an emollient for the hair and external layers of epidermal cells as well as serving as a bactericide. **(Glands)**

17. Nails are composed of tightly packed, keratinized cells. The nail body is the visible portion with a free edge which may extend past the distal end of the digit. The nail root extends into a fold of skin and is continuous with the stratum basale of the epidermis. A thickened area of the nail bed, called the nail matrix is responsible for growth. The region of the nail body that overlies the nail matrix appears as a white crescent called the lunula. Nails protect the end of the digits and serve as "tools" for the manipulation of small objects. **(Epidermal derivatives)**

18. e **(Epidermal derivatives)**

19. a **(Epidermal derivatives)**

20. h **(Epidermal derivatives)**

21. d **(Epidermal derivatives)**

22. b **(Epidermal derivatives)**

23. f **(Epidermal derivatives)**

24. g **(Epidermal derivatives)**

25. c **(Epidermal derivatives)**

26. The skin produces thermal regulation in two principal ways. The perspiration secreted by sweat glands evaporates to lower the epidermal temperature. Dilation or constriction of the blood flow to the capillary nets of the papillary layer will alter the amount of heat loss from the blood to the external environment. **(Functions of the skin)**

27. The protective functions of the skin include the chemical, physical, and biological barriers. The chemical secretion of sebum serves as a bactericide and the production of melanin serves to absorb UV radiation. The physical overlapping, layered structure of stratified squamous epithelium, like the shingles of a roof, reduce the likelihood of infection while effectively blocking the diffusion of water and water-soluble substances. In addition, these keratinized cells are resistant to the wear and tear from the environment. The Langerhans cells and macrophages found in the epidermis and dermis are active elements in the biological immune system. **(Functions of the skin)**

28. The skin may contain 200 sensory receptors per square centimeter. For this reason, it is virtually impossible to find an area of the skin that is insensitive to sensations of touch, pressure, heat, cold, pain, or vibration. **(Functions of the skin)**

29. The skin excretes small amounts of ions and small organic molecules as components of sweat. The synthesis of vitamin D begins in the cells of the skin that utilize the energy of UV radiation to modify a precursor cholesterol molecule. This molecule is further processed in the liver and kidneys to produce calcitrol, the most active form of vitamin D. **(Functions of the skin)**

30. The hypodermis is also called the subcutaneous layer and is composed primarily of loose connective tissues such as areolar and adipose. Approximately half the body's supply of adipose tissue is found in the subcutaneous fat. It is an excellent insulator and shock absorber and anchors the skin to the organs below. **(Hypodermis)**

31. Deep wound healing begins with the inflammatory phase in which a blood clot begins to form. Vasodilation and increased permeability deliver white blood cells such as macrophages and neutrophils that serve to phagocytize invading microbes. In addition, mesenchymal cells are delivered which develop into fibroblasts. In the migratory phase, the clot becomes a scab as epithelial cells migrate beneath the scab to bridge the wound. Fibroblasts begin to synthesize the scar tissue (collagen fibers), and the wound fills during the granulation phase. The proliferation phase is characterized by growth of epithelium and growth of blood vessels. The healing process ends with the maturation phase in which the scab is sloughed off as the epithelium is restored to near normal organization. **(Skin homeostasis)**

32. f **(Skin)**

33. e **(Skin)**

34. b **(Skin)**

35. h **(Skin)**

36. d **(Skin)**

37. g **(Skin)**

38. a **(Skin)**

39. c **(Skin)**

Grade Yourself

Circle the numbers of the questions you missed, then fill in the total incorrect for each topic. If you answered more than three questions incorrectly, you need to focus on that topic. (If a topic has less than three questions and you had at least one wrong, we suggest you study that topic also. Read your textbook, a review book, or ask your teacher for help.)

Subject: Integumentary System

Topic	Question Numbers	Number Incorrect
Skin	1, 32, 33, 34, 35, 36, 37, 38, 39	
Epidermis	2, 3, 4, 5, 6, 7	
Dermis	8, 9, 10, 11, 12, 13, 14	
Glands	15, 16	
Epidermal derivatives	17, 18, 19, 20, 21, 22, 23, 24, 25	
Functions of the skin	26, 27, 28, 29	
Hypodermis	30	
Skin homeostasis	31	

Skeletal System and Articulations

6

Brief Yourself

The internal skeletal system is a marvel of nature's engineering. Not only does the skeletal system provide a rigid, support framework for the soft tissues of the body to attach, it provides for the capability of rapid movement by producing a series of levers through the articulations that exist between its rigid components. In addition, the skeletal system serves the physiological functions of storing certain minerals and producing the cells of the blood.

The organs that comprise the skeletal system are the bones and their associated tissues and articulations (joints). The largest component of bone is the osseous tissue, a specialization of connective tissue. This specialized connective tissue produces a unique matrix that provides for bones that have both a degree of rigidity and a degree of flexibility. This combination allows for bones to withstand great stresses prior to reaching the breaking point. Osseous tissue is a particularly active tissue. Through the growth and reabsorption of osseous tissue, bones can be constantly remodelled throughout life.

The bones of the skeletal system are joined together at the articulations (joints). These articulations vary from being immovable, such as in the sutural joints of the skull, to freely movable, such as in the ball and socket joint of the shoulder. This variety of different types of articulations illustrates the varied mechanical functions of the skeletal system.

Test Yourself

1. List and describe the mechanical functions of bone. List and describe the physiological functions of bones.

2. Label the diagram of the long bone below.

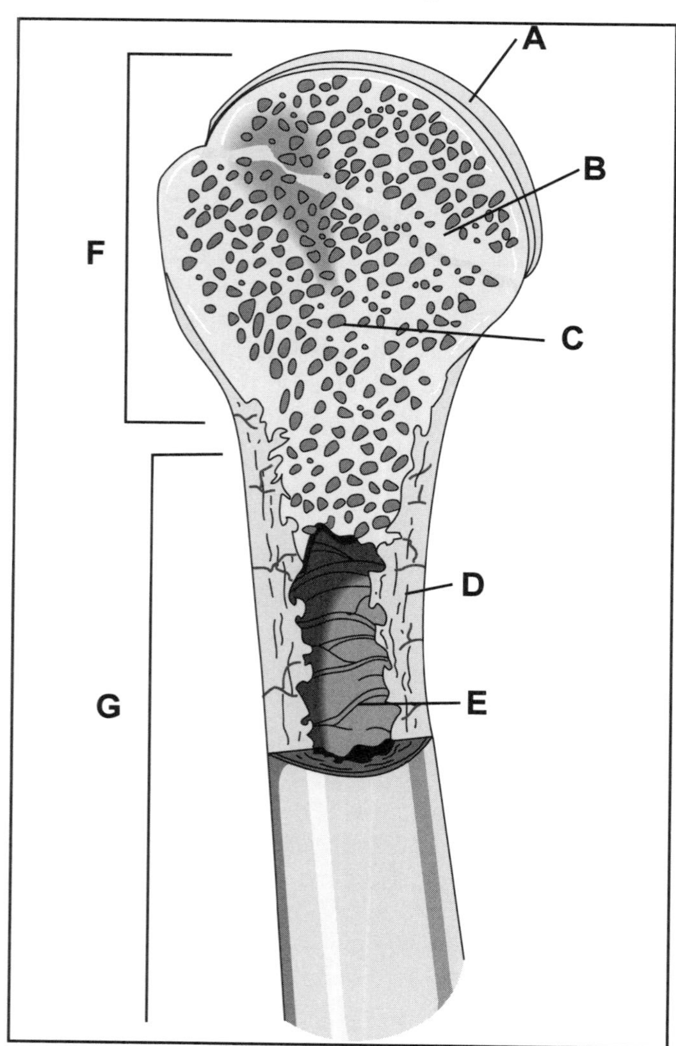

Fig 6-1

3. Identify and give the functions of the four types of bone cells.

4. What is bone matrix? How does it differ from the matrix typically produced by connective tissues?

5. How does spongy (cancellous) and compact (cortical) bone differ in structure and function?

6. Label the diagram of the bone section below.

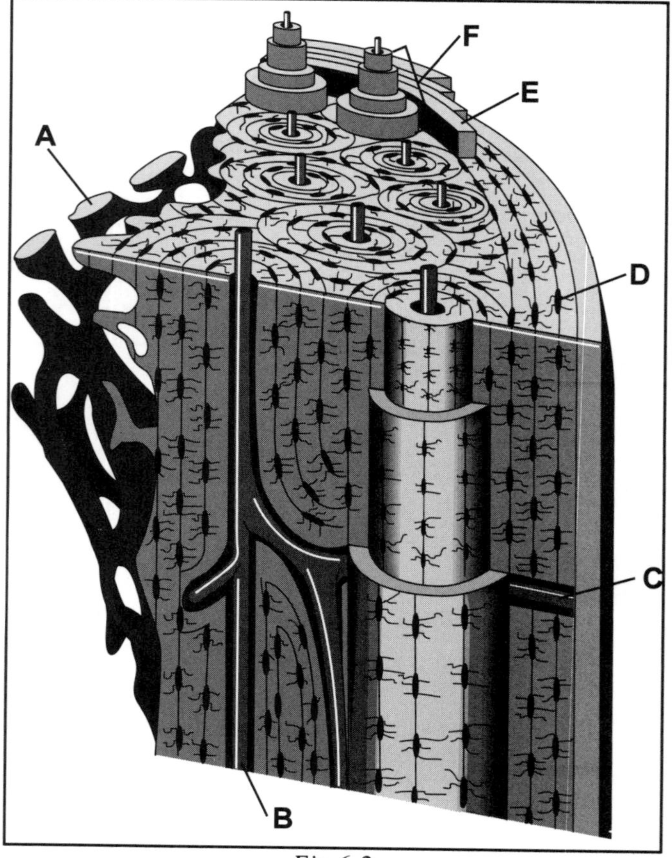

Fig 6-2

7. What are the two methods by which a bone is formed? What are the principal differences in these two methods?

8. List and describe the stages of endochondral ossification.

9. What is the epiphyseal plate (disk) and how does it allow for bone elongation? What is the epiphyseal line?

10. What is appositional bone growth and what is its significance?

11. Describe the role of bone tissue in the maintenance of calcium homeostasis. Identify the hormones involved.

Question 12–19 are matching. Match the type of fracture with its description.

12. Bone splintered at the site of impact with small fragments interspaced between the main fragments
13. Bone broken into two or more pieces
14. Bone broken at a point of weakness brought about by the effects of disease
15. Bone partially fractured with one side of the bone broken and other side of the bone dramatically bent
16. Broken bone does not break through the skin
17. Microscopic fractures typically produced by repeated strenuous activity
18. Broken bone protrudes through the skin
19. Ends of the broken bones are driven into one another

a. Complete
b. Simple
c. Compound
d. Greenstick
e. Impacted
f. Stress
g. Comminuted
h. Pathologic

20. Identify and describe the four principal types of bones.

Questions 21–32 are matching. Match the following bone marking with its description.

21. A large, rounded articular prominence
22. A large, rounded and roughened process
23. A tube-like passageway through a bone
24. Air filled cavity within a bone
25. Cleft-like opening between bones, typically for the passage of blood vessels or nerves
26. A rounded articular structure supported by a constricted portion (neck) of a bone
27. A depression on a bone
28. A small, rounded process
29. A smooth, flat surface
30. An opening through which blood vessels or nerves pass
31. A large projection found on the femur
32. A prominence superior to a condyle

a. Fissure
b. Foramen
c. Fossa
d. Meatus
e. Sinus
f. Condyle
g. Facet
h. Head
i. Epicondyle
j. Trochanter
k. Tubercle
l. Tuberosity

33. What are the two main divisions of the vertebrate skeleton? How might each be further sub-divided?

34. The skull is comprised of cranial and facial bones. How are the two groups differentiated? List the bones that comprise the cranium. List the facial bones.

35. What is a suture? Identify the four primary sutures.

36. What are the four types of vertebrae? What are the characteristics of each group that serve to distinguish them?

37. How are ribs classified?

38. Identify the bones of the pectoral girdle and describe its function.

39. List the bones of the upper limb (appendage).

40. List the bones of the pelvic girdle and describe its function.

41. List the bones of the lower limb (appendage).

42. What are the principal differences between the male and female pelvis? What other differences can be expected between male and female skeletons?

43. What is an articulation? What are the classifications of joints based upon their structure and function?

44. List and describe the three types of synarthrotic joints and the two types of amphiarthrotic joints.

45. Describe the structures of a typical diarthrotic (synovial) joint.

46. What are accessory ligaments and articular discs found in some synovial joints?

47. List the six types of diarthrotic joints and give an example of each.

48. Movements are said to be of four general types; gliding, rotation, angular, and special. Identify the six angular movements.

49. Identify the ten special movements and the joints where these movements occur.

50. How does the mobility of a joint relate to its structural stability?

Check Yourself

1. Mechanically, bones provide protection for certain organs, a rigid framework for supporting the soft tissues, and a system of levers to allow for movement. Physiologically, bones serve to help maintain mineral homeostasis by storing minerals, especially calcium and phosphorus, produce blood cells in the red bone marrow (hematopoiesis), and store some energy in the yellow bone marrow. **(Function of bone)**

2. a. Articular cartilage

   b. Epiphyseal line

   c. Spongy bone

   d. Cortical bone

   e. Medullary cavity

   f. Proximal ephiphysis

   g. Diaphysis **(Structure of bone)**

3. Bone cells include:

   a. Osteoprogenitor—unspecialized mesenchymal cells that can develop into osteoblasts.

   b. Osteoblasts—cells that form osseous matrix by secreting collagen and mineral salts.

   c. Osteocytes—developed from osteoblasts that have become trapped in the matrix. These cells maintain existing osseous matrix.

   d. Osteoclasts—function in the reabsorption of the osseous matrix. **(Microscopic structure of bone)**

4. The matrix of bone is similar to other connective tissues in that it is produced primarily of heavy collagen fibers. This tough, structural protein provides the tensile strength to bones. Bone matrix differs from other connective tissues in its rigidity (hardness). The rigidity is produced by the calcification (mineralization) of bone matrix due to the secretion of mineral salts by the bone producing cells (osteoblasts). These calcium based salts, primarily hydroxyapatite and calcium carbonate, accumulate in the spaces between the collagen fibers and crystallize producing the rigidity of bone. **(Microscopic structure of bone)**

5. Compact bone appears to be very organized, consisting of cylinders of bone growth around a central canal containing blood vessels. Each cylinder of bone growth is called an osteon (Haversian system). Compact bone is particulary dense and rigid, providing a strong external layer in all bones and the bulk of the diaphysis in long bones. Spongy bone appears to grow in irregular patterns or plates called trabeculae leaving many open spaces. These spaces are filled with red bone marrow, a soft hematopoietic tissue. The inclusion of these macroscopic spaces provides a "spongy" quality to this matrix. Spongy bone makes up the majority of the tissue of short, irregular, and flat bones as well as most of the epiphyses of the long bones. **(Microscopic structure of bone)**

6. a. Spongy bone

   b. Central (Haversian) canals

   c. Perforating (Volkmann's) canals

   d. Osteocyte

   e. Lamellae

   f. Osteon **(Microscopic structure of bone)**

7. Bones are formed by ossification, which may be intramembranous or endochondral. Intramembranous ossification will produce most of the flat and irregular bones of the body while endochondral ossification will produce the long and short bones. Intramembranous ossification occurs as the replacement of fibrous connective tissue membranes. Endochondral ossification occurs as the replacement of cartilage. **(Bone formation)**

8. Bone osteoprogenitor cells begin to accumulate near the midpoint of the cartilage and begin to ossify the tissue forming a bone collar. The invagination of a blood vessel into the developing cavity (cavitation) beneath the collar is called the periosteal bud which brings about the primary ossification center as the capillaries induce growth of osteoblasts to replace the cartilage. The invagination of additional blood vessels develops secondary centers of ossification at the ends of the bones producing pads of cartilage between the primary and secondary center which become the epiphyseal disks, resulting in the mature form of the bone. **(Bone formation)**

9. The epiphyseal disk is the growth plate in long bones that is found between the two epiphyses and the diaphysis. The plate is composed primarily of hyaline cartilage. The cells closest to the epiphysis divide rapidly and the production of new cartilage matrix causes the elongation of the bone. The cells nearest the diaphysis are typically replaced at approximately the same rate. The decreased production of growth hormone at the end of adolescence results in a decreased production of somatomedin causing a slower growth of cartilage cells. The faster growing bone cells eventually replace all of the epiphyseal disk resulting in a "closure" of the growth plate producing an epiphyseal line. **(Bone growth)**

10. Appositional growth is the increase in the diameter of bones. Although elongation is limited by the closure of epiphyseal disks at maturity, the remodelling of bones due to stress alterations continues throughout life. Bones continue to grow in response to stresses by the production of new matrix from the osteoblasts beneath the periosteum. **(Bone growth)**

11. In the event that the calcium level in the blood begins to drop below acceptable levels, the parathyroid gland responds by the release of parathyroid hormone (parathormone) which acts to stimulate the activity of osteoclasts. The osteoclasts will reabsorb bone matrix, releasing the calcium to the blood stream raising the calcium concentration. In the event that the calcium level in the blood becomes too high, the C cells of the thyroid gland release calcitonin which inhibits the activity of osteoclasts (and thus decreases the release of calcium from the matrix) **(Calcium homeostasis)**

12. g **(Bone fractures)**

13. a **(Bone fractures)**

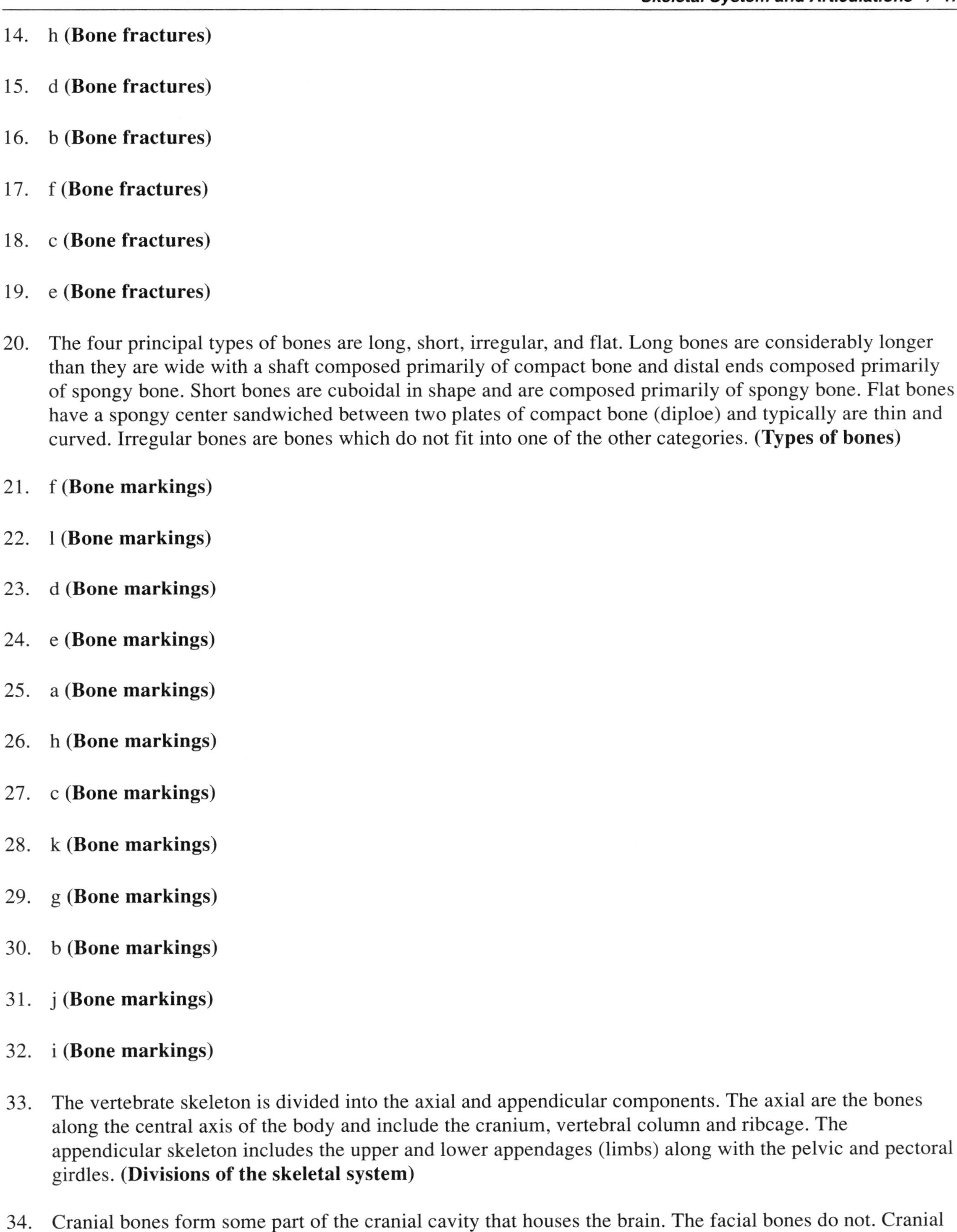

14. h **(Bone fractures)**

15. d **(Bone fractures)**

16. b **(Bone fractures)**

17. f **(Bone fractures)**

18. c **(Bone fractures)**

19. e **(Bone fractures)**

20. The four principal types of bones are long, short, irregular, and flat. Long bones are considerably longer than they are wide with a shaft composed primarily of compact bone and distal ends composed primarily of spongy bone. Short bones are cuboidal in shape and are composed primarily of spongy bone. Flat bones have a spongy center sandwiched between two plates of compact bone (diploe) and typically are thin and curved. Irregular bones are bones which do not fit into one of the other categories. **(Types of bones)**

21. f **(Bone markings)**

22. l **(Bone markings)**

23. d **(Bone markings)**

24. e **(Bone markings)**

25. a **(Bone markings)**

26. h **(Bone markings)**

27. c **(Bone markings)**

28. k **(Bone markings)**

29. g **(Bone markings)**

30. b **(Bone markings)**

31. j **(Bone markings)**

32. i **(Bone markings)**

33. The vertebrate skeleton is divided into the axial and appendicular components. The axial are the bones along the central axis of the body and include the cranium, vertebral column and ribcage. The appendicular skeleton includes the upper and lower appendages (limbs) along with the pelvic and pectoral girdles. **(Divisions of the skeletal system)**

34. Cranial bones form some part of the cranial cavity that houses the brain. The facial bones do not. Cranial bones include: frontal, parietal (2), temporal (2), occipital, sphenoid, and ethmoid. The facial bones

include; nasal (2), maxillae (2), mandible, lacrimal (2), palatine (2), inferior nasal conchae (2), vomer, and zygomatic (2). **(Bones of the skull)**

35. A suture is a fixed joint between the cranial bones. The four primary sutures are the coronal, sagittal, lambdoid, and squamous. **(Bones of the skull)**

36. The vertebral column is divided into cervical, thoracic, lumbar, and sacral vertebrae (along with the vestigial coccygeal). The cervical are the smallest in mass and include two features found in no other group; transverse foramen and bifurcated spiny processes. The thoracic are larger in mass and include costal facets (the points of rib articulation). The lumbar are the largest vertebrae with a massive body. The sacral vertebrae are fused into a single plate called the sacrum. **(Vertebral column)**

37. Ribs are classified according to their attachment to the sternum. The first seven pairs of ribs attach directly to the sternum and are called true ribs. Ribs eight, nine, and ten have cartilage that attaches to the cartilage of the seventh rib and are called false ribs. The eleventh and twelth ribs do not attach to the sternum and are called floating ribs. **(Ribcage)**

38. The pectoral girdle consists of the clavicle and scapula. Its function is to serve as the site of attachment of the upper limbs. The clavicle has two articulations: the sternoclavicular joining it to the sternum of the axial skeleton, and the acromioclavicular joining it to the acromion of the scapula. The scapula forms the humeroscapular joint for the point of attachment to the upper limb. **(Upper limb)**

39. Each upper limb consists of a humerus, radius, ulna, eight carpals, five metacarpals, and fourteen phalanges. **(Upper limb)**

40. The pelvic girdle consists of two coxal bones which form a common joint at the pubic symphysis. The coxal bone is comprised of three fused bones, the ilium, ischium, and pubis. In addition to the joint at the pubic symphysis, the coxal bone forms two more articulations: the sacroiliac joint for attachment to the axial skeleton and the coxafemoral joint for attachment to the lower limb. **(Pelvic girdle)**

41. Each lower limb consists of a femur, tibia, fibula, patella, seven tarsals, five metatarsals, and fourteen phalanges. **(Lower limb)**

42. The female pelvis is specialized for pregancy and childbirth. The female pelvis is more shallow and more oval than the male pelvis. It is slightly larger with a pubic arch angle of greater than 90 degrees. It is less vertical and produces a slightly greater angle of attachment for the femurs. Due to the larger and more massive muscular development in males, the male skeleton tends to be larger and heavier with more prominent bone markings at points of tendon attachment. **(Sex differences in the skeletal system)**

43. An articulation is the point of contact between two or more bones (or between bone and cartilage or teeth). Functionally, joints are classified as synarthrotic, which typically are immovable, amphiarthrotic which are slightly movable, and diarthrotic, which are freely movable. **(Classification of joints)**

44. The synarthrotic joints include the synostosis, gomphosis, and synchondrosis. Synostoses are joints connected by a thin fibrous connective sheath that often are completely replaced by bone such as a suture. Gomphoses are specialized fibrous joints such as the articulations between the mandible or maxillae and the teeth. Synchondroses are joints connected by cartilage such as the articulation of the ribs to the sternum or within a long bones as the diaphysis connects to the epiphysis by the growth plate. The amphiarthrotic joints include the syndesmoses and the symphyses. The syndesmoses are joints connected by a relatively thick fibrous connective tissue sheath allowing some movement such as in the joints

between the shafts of tibia and fibula. The symphyses are fibrocartilagenous joints such as that found between the vertebrae. **(Classification of joints)**

45. A synovial joint consists of an articular capsule that completely encapsulates the ends of the articulating bones. The capsule is composed of two layers, the fibrous capsule and the synovial membrane. The external fibrous layer is primarily dense, irregular connective tissue and includes the external ligaments of the joint. The internal synovial membrane is composed primarily of specialized areolar connective tissue that secretes synovial fluid which fills the synovial cavity, or space inside the capsule. The ends of articulating bones in synovial joints are covered with articular cartilage (hyaline). The synovial fluid acts to lubricate the joint and provides nutrients to the chondrocytes of the articular cartilage. The hyaline cartilage serves as an excellent shock absorber and provides a low friction surface for these freely movable joints. **(Structure of synovial joints)**

46. Accessory ligaments are ligaments that help maintain joint stability but are not continuous with the articular capsule, either extracapsular or intracapsular. Examples of both are found in the knee joint, with the cruciate ligaments found inside the joint capsule and the popliteal found outside. Articular discs (menisci) are fibrocartilage pads attached to the bone surfaces (atop the hyaline cartilage) that act to provide additional shock absorption and channel the synovial fluid to the sites of the greatest friction. **(Structure of synovial joints)**

47. Diarthrotic joints include the gliding (intercarpal), hinge (humeroulnar), pivot (axoatlantic), condyloid (radiocarpal), saddle (carpo-metacarpal of thumb), and ball and socket (humeroscapular). **(Classification of joints)**

48. Angular movements include extension, flexion, hyperextension, circumduction, abduction, and adduction. **(Classification of movements)**

49. The ten special movements are those that occur only at particular joints. At the ankle, these movements include plantar flexion, dorsiflexion, inversion, and eversion. At the shoulder girdle and mandibular joint, these include protraction, retraction, elevation, and depression. At the elbow, supination and pronation is possible. **(Classification of movements)**

50. As a general rule, the greater the mobility of the joint the less structurally stable it becomes. Mobility is measured as the number of directions (planes) of movement available at a joint as well as the range of movement in each direction. Mobility is limited by shape of the articulation surfaces of the bones as well as by the soft tissues, such as the arrangement of ligaments and tendons surrounding the joint. Typically, the more limiting the anatomical structure, the more stable the joint becomes, and thus, the less mobile. **(Structure of synovial joints)**

Grade Yourself

Circle the numbers of the questions you missed, then fill in the total incorrect for each topic. If you answered more than three questions incorrectly, you need to focus on that topic. (If a topic has less than three questions and you had at least one wrong, we suggest you study that topic also. Read your textbook, a review book, or ask your teacher for help.)

Subject: Skeletal System and Articulations

Topic	Question Numbers	Number Incorrect
Function of bone	1	
Structure of bone	2	
Microscopic structure of bone	3, 4, 5, 6	
Bone formation	7, 8	
Bone growth	9, 10	
Calcium homeostasis	11	
Bone fractures	12, 13, 14, 15, 16, 17, 18, 19	
Types of bones	20	
Bone markings	21, 22, 23, 24, 25, 26, 27, 28, 29, 30, 31, 32	
Divisions of the skeletal system	33	
Bones of the skull	34, 35	
Vertebral column	36	
Ribcage	37	
Upper limb	38, 39	
Pelvic girdle	40	
Lower limb	41	
Sex differences in the skeletal system	42	
Classification of joints	43, 44, 47	
Structure of synovial joints	45, 46, 50	
Classification of movements	48, 49	

Muscular System

Brief Yourself

One clear indication of the importance of muscles in human anatomy and physiology is its amount. In all of its forms, muscle makes up almost half of the mass of a human and is the dominant tissue of the musculoskeletal system, the cardiovascular system, and the walls of all the hollow organs of the body.

The most critical physiological characteristic of muscle is the ability of its cells to transform the chemical energy of ATP into a mechanical energy and thus, the production of force. These forces are utilized to produce mobility for the body as a whole or in part, as well as providing for the movement of fluids or other substances through the internal structures of the body. Muscle tissues also function in the stabilization of body position, regulation of organ volumes, and the generation of heat as a byproduct of its metabolism.

Test Yourself

1. List the three types of muscle tissue and identify the general location, anatomical characteristics, and nervous (or endocrine) control differences among them.

2. List and briefly describe the four primary functions of muscle tissue.

3. List and briefly describe the five principle characteristics of muscle tissue required for its functional capabilities.

Questions 4–17 are matching. Match the following with its description.

4. The sole chemical messenger used at the neuromuscular junction of skeletal muscles — a. Deep fascia

5. A broad, flat layer of connective tissue that typically connects skeletal muscle to bone — b. Epimysium

6. A bundle of between 10 and 100 individual muscle fibers — c. Perimysium

7. A single motor neuron and all of the skeletal muscle fibers it innervates — d. Fascicle

8. Dense, irregular connective tissue that holds muscles together and separates them into functional groups	e. Endomysium	14. Penetrating from the epimysium, a layer of fascia that surrounds fascicles	k. Neurotransmitter
9. A specialized region of plasma membrane and the space between them that serves as a point of communication between excitable cells	f. Tendon	15. A specialized area of muscle fiber plasma membrane containing acetylcholine receptors	l. Myoneural junction
10. A general term for the chemical released by excitable cells which will alter the electrical potentials of the target cells	g. Aponeurosis	16. Penetrating from the perimysium, a layer of fascia that surrounds the individual muscle fibers	m. Acetylcholine
11. The outermost layer of fascia surrounding a whole muscle	h. Motor neuron	17. A neuron that innervates a muscle fiber	n. Motor end plate
12. A cord-like structure of dense connective tissue that connects a skeletal muscle to a bone	i. Motor unit		
13. A synapse between a motor neuron and a skeletal muscle fiber	j. Synapse		

18. Label the diagram of the microscopic view of the skeletal muscle below.

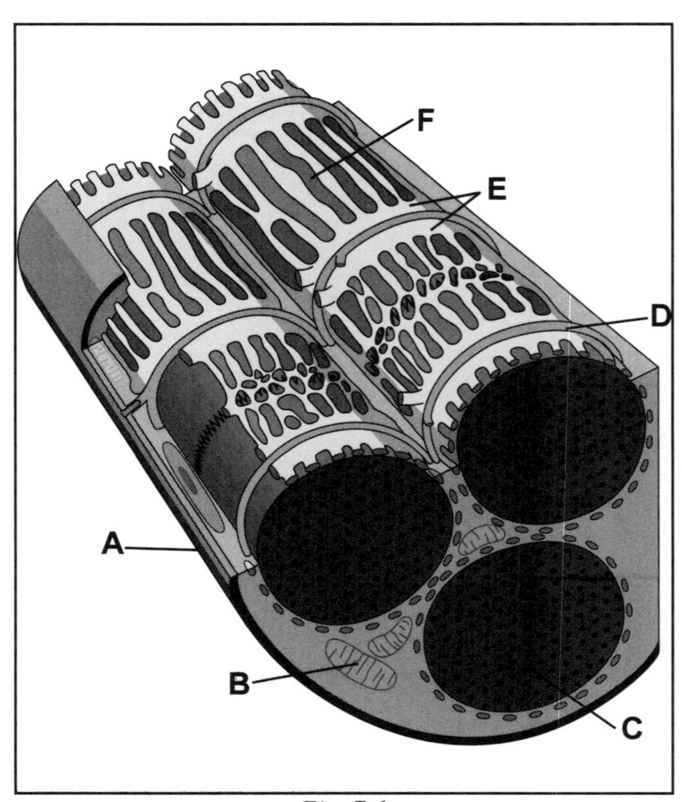

Fig. 7-1

19. Label the diagram of the sarcomere and myofilaments below.

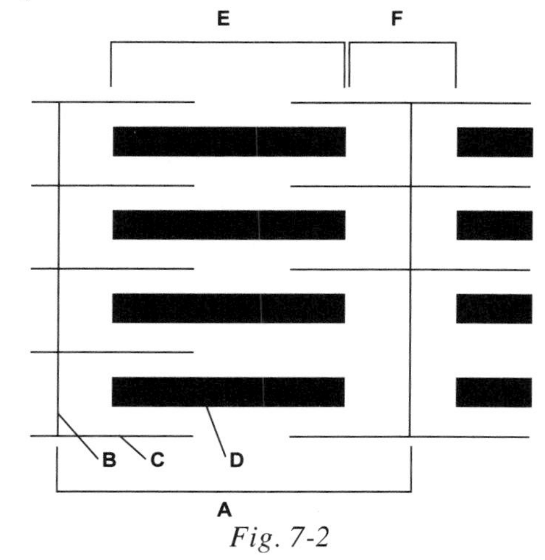

Fig. 7-2

20. What is excitation-contraction coupling?

21. What is a resting membrane potential? How is it produced in muscle fibers?

22. What is a muscle action potential? How are its effects carried from the sarcolemma into the sarcoplasm?

23. What is the role of calcium in muscle contractions? What is the relationship between calcium and the sarcoplasmic reticulum?

24. How does calcium release promote the production of crossbridges between actin and myosin?

25. What is the role of ATP in muscle contractions?

26. How does a muscle relax?

27. What is a muscle twitch? What does a myogram indicate about the nature of a muscle twitch?

28. Define and describe the following:

    a. Refractory period.

    b. Temporal summation.

    c. Tetanus.

    d. Treppe.

29. What is the length-tension relationship in muscle fibers?

30. How is the force produced by a muscle controlled?

31. What is the difference between isotonic and isometric contractions? What two forms of isotonic contractions exist?

32. What are the three metabolic systems by which muscle fibers produce ATP? How do they differ?

33. What produces muscle fatigue? What produces muscle soreness following intense exercise bouts?

34. List and describe the three muscle fiber types.

35. Describe cardiac muscle tissue with regard to its similarities and differences to skeletal muscle.

36. What are the two types of smooth muscle tissue and how do they differ?

37. Describe the microscopic anatomy of smooth muscle.

38. What is meant by the origin, insertion, and action of a muscle?

39. What is a lever and a fulcrum? Describe and give an example of the three classes of levers produced in the musculoskeletal system.

40. Define the following:

    a. Agonist.

    b. Antagonist.

    c. Synergist

    d. Fixator

41. List and describe the possible arrangements of fascicles in muscles. What is the functional significance of these arrangements?

Check Yourself

1. The three types of muscle tissue include:

    a. Skeletal, which is primarily attached to the skeletal system, appears striated, and is under voluntary nervous control.

    b. Cardiac, which forms most of the heart, appears striated (but not as distinctly as skeletal), and is under involuntary nervous and endocrine control.

    c. Smooth, which is located in the walls of hollow organs and in the skin, appears smooth or nonstriated, and is under involuntary nervous and endocrine control. **(Types of muscle tissue)**

2. The four primary functions of muscle tissue are:

    a. Motion produced by the integrated actions of the bones, joints, and muscles.

    b. Movement of substances through the body, such as the contractions of the cardiac tissue to pump blood through the circulatory system, or contractions of the smooth muscle in the walls of the gastrointestinal tract to move ingested material through the body.

    c. Stabilization of the position of the body and regulation of organ volumes, such as the contractions associated with posture or the utilization of sphincter muscles to control emptying of organs such as the stomach or urinary bladder.

    d. Generation of heat due to cellular respiration. **(Types of muscle tissue)**

3. The five characteristics of muscle necessary for its functions are:

    a. Excitability, the ability to respond to stimuli by the production of electrical potentials.

    b. Conductivity, the ability to propagate electrical potentials along the plasma membrane of its cells.

    c. Contractility, the ability to forcibly shorten.

    d. Extensibility, the ability to lengthen (stretch) without damage.

    e. Elasticity, the ability to return to an original length following contractions or extensions. **(Characteristics of muscle tissue)**

4. m **(Anatomy of muscle tissue)**

5. g **(Anatomy of muscle tissue)**

6. d **(Anatomy of muscle tissue)**

7. i **(Anatomy of muscle tissue)**

8. a **(Anatomy of muscle tissue)**

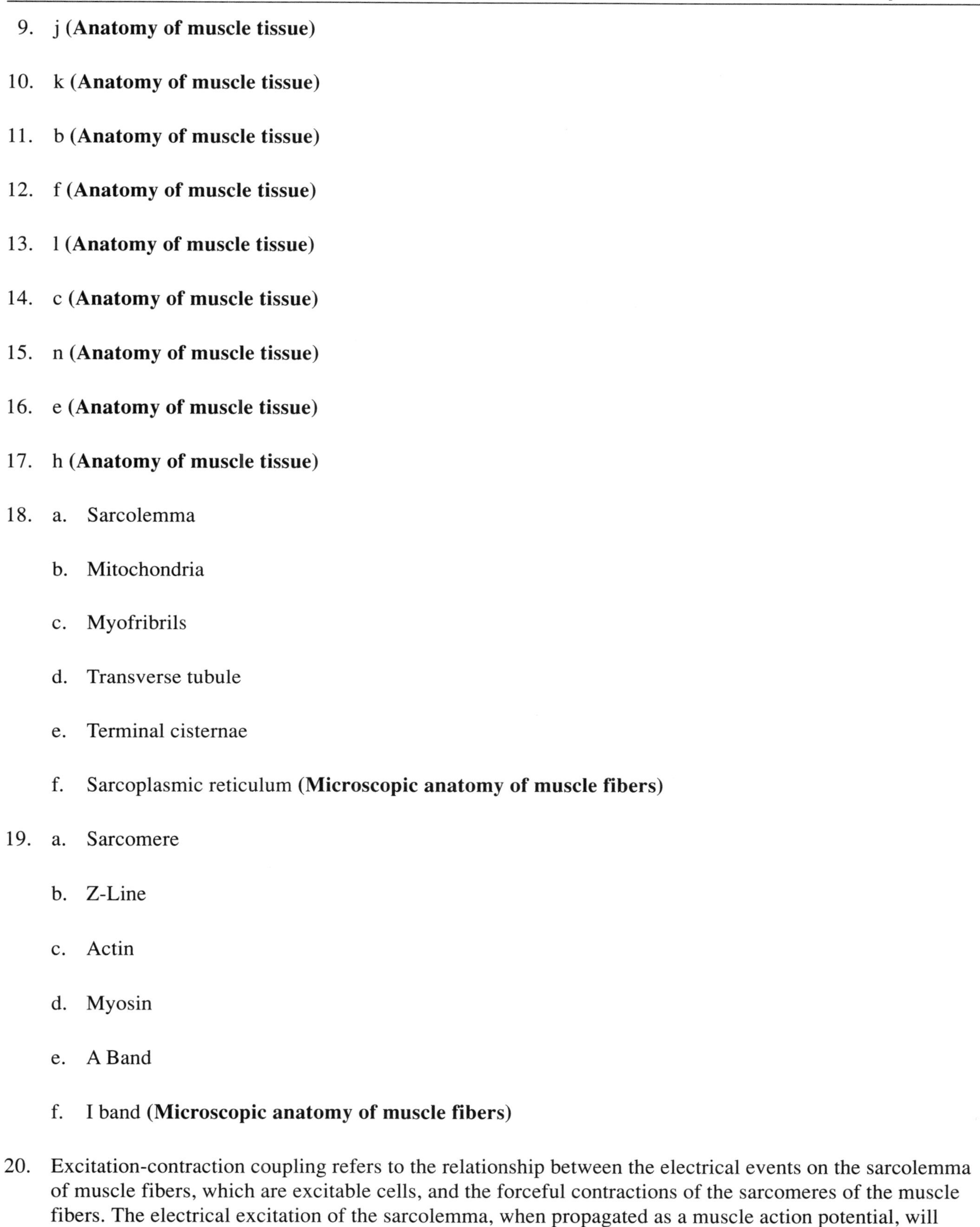

9. j **(Anatomy of muscle tissue)**

10. k **(Anatomy of muscle tissue)**

11. b **(Anatomy of muscle tissue)**

12. f **(Anatomy of muscle tissue)**

13. l **(Anatomy of muscle tissue)**

14. c **(Anatomy of muscle tissue)**

15. n **(Anatomy of muscle tissue)**

16. e **(Anatomy of muscle tissue)**

17. h **(Anatomy of muscle tissue)**

18. a. Sarcolemma

    b. Mitochondria

    c. Myofribrils

    d. Transverse tubule

    e. Terminal cisternae

    f. Sarcoplasmic reticulum **(Microscopic anatomy of muscle fibers)**

19. a. Sarcomere

    b. Z-Line

    c. Actin

    d. Myosin

    e. A Band

    f. I band **(Microscopic anatomy of muscle fibers)**

20. Excitation-contraction coupling refers to the relationship between the electrical events on the sarcolemma of muscle fibers, which are excitable cells, and the forceful contractions of the sarcomeres of the muscle fibers. The electrical excitation of the sarcolemma, when propagated as a muscle action potential, will trigger events inside the cell which will result in the development of contractions. **(Contraction of muscles)**

21. The resting membrane potential is an electrical potential that exists across the plasma membrane of cells. In muscle fibers, the sarcolemma is a selectively permeable membrane. Primarily through the utilization of membrane channels and sodium/potassium membrane pumps, considerable differences in the concentrations of cations and anions can be produced between the external and internal fluid compartments (extracellular vs. intracellular) resulting in a measurable voltage or a polarized membrane. **(Contraction of muscles)**

22. A muscle action potential is a rapid change in the resting membrane potential, called a depolarization, that can be propagated along the cell membrane, in this case the sarcolemma. It is produced by the release of acetylcholine from the innervating motor neuron. This neurotransmitter is released into the synaptic cleft of the myoneural junction and diffuses across to waiting acetylcholine receptors at the motor end plate. The binding of acetylcholine at the motor end plate opens chemically-mediated sodium channels, resulting in a rapid flow of sodium across the sarcolemma as it follows its chemical and electrical gradients. This ionic current opens voltage-mediated sodium channels along the length of the sarcolemma, propagating the potential all along the length of the muscle fiber. This electrical event is carried into the muscle fiber along the T-tubules, which are invaginations of the sarcolemma that run deep into the cell. **(Contraction of muscles)**

23. Calcium concentrations in the sarcoplasm are responsible for the movement of the filaments. Increases in the calcium concentration in the sacroplasm starts the filaments sliding, decreases turns them off. When a muscle is relaxed (not contracting) there is a relatively low calcium concentration in the sarcoplasm because the sarcoplasmic reticulum (SR) membrane contains calcium pumps that actively move calcium from the sarcoplasm into the SR utilizing ATP to power these pumps. Calcium is sequestered primarily in the terminal cisternae near the transverse (T) tubules. The cisternae of the SR contain a protein called calsequestrin which acts as a binding site for calcium. **(Contraction of muscles)**

24. The release of calcium from the SR occurs as the muscle action potential is propagated along the sarcolemma and into the T-tubule system, opening voltage-gated calcium channels in the SR membrane. The calcium binds to troponin, altering the shape (denaturing) the troponin-tropomyosin complex. This uncovers the binding sites on the actin molecules to allow for the chemical bonding with the globular heads of the myosin molecules producing actomyosin. **(Contraction of muscles)**

25. ATP acts to "power" the contraction. The physical shortening (or production of mechanical force) actually occurs due to shape changes in the myosin heads as they form, break, and reform bonds to the actin filament. The myosin heads contain binding sites for ATP, and when ATP is available and bound to the myosin head it is said to be in a high energy configuration. Immediately upon the binding of the high energy myosin head to actin, the ATP breaks down to ADP and an inorganic phosphate. As these particles are released, the myosin head changes shape, producing a "ratcheting" movement or power stroke as the head swivels toward the center of the sarcomere. This action draws the actin filament past the myosin filament toward the H zone, effectively producing a mechanical shortening (force). The low energy myosin head will attract and bind to another ATP molecule, attach to a new binding site on actin and the process of shortening continues. **(Contraction of muscles)**

26. Relaxation occurs with the cessation of the muscle action potential. Acetylcholinesterase in the synaptic cleft of the myoneural junction rapidly degrades acetylcholine, causing its effect to be short-lived. The muscle action potential ends, the membrane pumps of both the sarcolemma and SR membrane actively move ions to re-establish the resting membrane potential across the sarcolemma and to actively sequester calcium in the SR. With the calcium concentration in the sarcomere reduced, the troponin-tropomyosin complex regains its resting shape, covering the binding sites on actin and effectively preventing crossbridge formation. The sarcomeres then returns to their resting lengths. **(Contraction of muscles)**

27. A muscle twitch is a contraction of all the muscle fibers in any single motor unit in response to a single action potential from its motorneuron. A graphic record of this response is a myogram. A myogram will show three components:

    a. The latent period, which shows a slight delay between the delivery of the stimulus and the production of appreciable force. This is due to the time necessary for the delivery of calcium from the SR and time necessary to take the "slack" out of the myoelastic component of the muscle.

    b. The contraction period, in which forces increase as the sarcomeres attempt to shorten.

    c. The relaxation period, in which forces decrease following the cessation of the stimulus and the resetting of the muscle resting potential and the sequestering of calcium. **(Muscle tension production)**

28. a. Refractory period is a brief period of lost excitability in excitable cells due to the time required to re-establish necessary membrane potentials to propagate impulses. In muscle fibers it varies from about 5 msecs in skeletal muscle to 300 msecs in cardiac muscle.

    b. Temporal summation is the phenomenon of a differential response to two identical stimuli applied in rapid succession. If a stimulus is applied, followed by a second identical stimulus that is applied quickly after the first, but delayed enough to occur following the refractory period, the response to the second stimulus will be greater than the first.

    c. Tetanus is a sustained muscle contraction that occurs when the stimuli are applied so rapidly that the muscle fibers cannot relax prior to the arrival of the next stimulus. In such a case, calcium is being released much more rapidly than it can be actively sequestered, resulting in continued contractions.

    d. Treppe is a "stairstep" increase in the force of contractions that occurs in muscle fibers with successive identical stimuli that are too far apart for temporal summation to be acting. Once again, although the stimuli are not applied rapidly enough for tetanus to occur, they are still being produced at a frequency that does not allow all the calcium to be sequestered prior to the next stimulus, resulting in slightly increased contractions with each stimulus applied. **(Frequency of stimulation)**

29. The length-tension relationship of a muscle is related to the overlap between the actin and myosin filaments. In order for contractions to occur, some overlap between these filaments is required. If a muscle were overstretched to the point that the myosin and actin did not overlap, no crossbridges could be formed and thus no contraction would be possible. A muscle will produce the maximal contraction force when an optimal overlap occurs at an optimal length. Increase or decreases in the optimal length will result in lowering the force of contraction. **(Length-tension relationship)**

30. The primary control of forces produced in a muscle is through motor unit recruitment. Muscles are composed of motor units, each consisting of a single motorneuron and all the muscle fibers that it innervates. Each motor unit will produce force as an "all or none" phenomenon. When activated, each muscle fiber will maximally contract. To alter the amount of contraction force in whole muscles, different numbers of motor units can be activated, recruiting additional motor units as greater and greater forces are required. **(Motor unit recruitment)**

31. Isotonic contractions occur when movement at a joint occurs. In other words, an isotonic contraction is one in which the force produced by the muscle is either greater or lesser than the external load placed on the muscle. Isotonic contractions can be either concentric, in which the force produced is greater than the external load such that the muscle will shorten, or eccentric, in which the force produced is less than the external load such that the muscle will lengthen. Isometric contractions are those in which no movement

occurs at a joint. In this case, the amount of force produced in the muscle must exactly equal its external load. **(Types of muscle contractions)**

32. The three metabolic systems that provide ATP for muscle contractions are the phosphagen system, the anaerobic system (glycogen-lactic acid), and the aerobic system. The phosphagen system provides energy when muscles are contracted maximally. This system utilizes stores of existing ATP and creatine phosphate (CP), a molecule that can transfer its high-energy phosphate unit directly to ADP. Together, ATP and CP can provide approximately 10-15 seconds of energy during its maximal use. The anaerobic system is also a rapid producer of energy, utilizing glycolysis to provide ATP. It is limited by the ability of the cell to tolerate lactic acid accumulation, a byproduct of these reactions. When energy metabolism occurs more rapidly than oxygen can be delivered to the cells, lactic acid accumulates. Typically, anaerobic metabolism can supply about 30-40 seconds of maximal muscle activity prior to the critical accumulation of acidity. The aerobic system utilizes cellular respiration and can provide theoretically unlimited energy provided that the pace of energy production can be matched by the delivery of oxygen, resulting in the production of pyruvic acid rather than lactic acid. Unlike lactic acid, pyruvic acid can be further metabolized by the mitochondria of the muscle fibers so that no appreciable acidity builds up. **(Muscle metabolism)**

33. Muscle fatigue occurs due to a number of factors, but truly results when there is insufficient ATP to meet the needs of contracting muscles. The factors include insufficient oxygen (which leads to the accumulation of lactic acid), insufficient glycogen (glucose), or failure of motor neurons to release acetylcholine. Delayed onset muscle soreness is often mistakenly associated with a buildup of lactic acid in the muscle. Approximately 80 percent of lactic acid produced in muscles diffuses out of the muscle immediately upon production. The remaining 20 percent is rapidly metabolized when muscles come to rest. Muscle soreness after intense bouts of exercise occurs due to the accumulation of fluid (swelling) in the muscle that results form microscopic damage to the myoelastic component of muscles during sustained large contractions. **(Muscle fatigue)**

34. The three types of muscle fibers are:

    a. Slow oxidative (type I) fibers which have the smallest diameter and contain large amounts of myoglobin and many mitochondria, and are well supplied with capillaries. They appear red in color and have the capacity to generate ATP primarily using the aerobic system, and as such, are very resistant to fatigue, although they are producing a slow contraction velocity.

    b. Fast oxidative-glycolytic (type IIa) fibers which have an intermediate diameter, large amounts of myoglobin, and many mitochondria, and are well supplied with capillaries. They appear red in color and have the capacity to generate ATP through both aerobic and anaerobic systems. For this reason, they produce a higher contraction velocity than the slow oxidative fiber but are not as resistant to fatigue.

    c. Fast glycolytic (type IIb) fibers which have the largest diameter, low myoglobin content, relatively few mitochondria, and relatively poor capillary supply. For this reason they appear white in color and generate the majority of their ATP by anaerobic systems; therefore, their contraction velocity is very high, but they are poorly resistant to fatigue. **(Muscle fiber types)**

35. Cardiac muscle fibers are striate, having the same basic arrangement of sarcomeres as found in skeletal muscle, but are shorter in length and larger in diameter. Unlike skeletal muscle fibers, cardiac muscle fibers exhibit branching and are uninucleate. They have a less extensive SR system than is found in skeletal muscle and as such must utilize extracellular calcium as part of their contraction process. The slower delivery and removal of calcium across the sarcolemma produces a prolonged contraction and refractory period. Cardiac cells utilize aerobic systems for the production of ATP almost exclusively and must be well supplied with

oxygen and thus have extensive capillary networks (from the coronary artery system). Unlike skeletal muscle fibers which are insulated from one another, cardiac muscle fibers are connected to one another at the ends of adjoining fibers by an intercalated disk of the sarcolemma. This contains many gap junctions which facilitate the spread of the muscle action potential from one fiber to the next, allowing for the spread of electrical excitation across the tissue. **(Cardiac muscle)**

36. Smooth muscle tissue is either visceral (single-unit) or multiunit. Visceral is the most common form, wrapped around the walls of hollow organs. The fibers of this muscle are connected by gap junctions to allow the spread of electrical excitation from fiber to fiber. Multiunit fibers are found in arrector pili muscles of hairs and in the large airways of the lungs and in large arteries. Each fiber has its own motor neuron and excitation typically results in the contraction of single units. **(Smooth muscle)**

37. Smooth muscle fibers are spindle shaped, have a central nucleus, and display no striations. Smooth muscle fibers do contain both thick and thin myofilaments, but they are not arranged in regular sarcomeres. In addition, smooth muscle fibers contain intermediate filaments attached to structures called dense bodies that are dispersed throughout the sarcoplasm as well as anchored to the sarcolemma. During contractions, the sliding action of the thick and thin filaments is transferred to the intermediate filaments and thus to the dense bodies attached to the sarcolemma causing a shortening of the distances between various portions of the sarcolemma. **(Smooth muscle)**

38. The origin of a muscle is the point of attachment of that muscle tendon to the relatively stationary (fixed) bone during contraction of the muscle. The insertion is the point of attachment to the relatively movable bone. The action of a muscle is the description of the type of movement that occurs at the joint it controls. **(Skeletal muscle movements)**

39. A lever is defined as a fixed rod that moves (rotates) over a fixed point called the fulcrum. A mechanical advantage, called leverage, is gained by the use of a lever. The musculoskeletal system is a system of levers in which bones act as levers, joints act as fulcrums, and muscles supply forces. First class levers have the fulcrum between the force and the resistance and are relatively rare in the body. An example of a first class lever would be extension at the atlanto-occipital joint. Second class levers have the fulcrum at one end with the force applied at other end and the resistance between them. Again, this is relatively rare with plantar flexion of the foot as an example. Third class levers have the fulcrum at one end and the resistance at the opposite end with force applied between. This is the most common lever of the body with flexion at the elbow as an example. **(Leverage)**

40. a. Agonist is a prime mover, the muscle that is most responsible for producing a desired movement.

    b. Antagonist is the opposite mover, the muscle that would be most responsible for producing an opposite movement.

    c. Synergist is a muscle that serves to aid or steady a desired movement.

    d. Fixator acts to stabilize the origin of prime movers so that the lever action of the prime mover can be utilized most efficiently. **(Skeletal muscle movements)**

41. Muscle fascicles can be arranged in the following patterns:

    a. Parallel, in which the fascicles are parallel to the longitudinal axis of the muscle and terminate in aponeuroses. They have relatively few fascicles distributed over the aponeurosis and thus produce greater range of motion than power.

b. Fusiform, in which fascicles are also nearly parallel to the longitudinal axis of the muscle but terminate in tendons, producing a "tapering" of the muscle from the thicker central belly to the tendons. This provides a larger number of fascicles distributed over the tendon and produces a balance of range of motion and power.

c. Circular, in which the fascicles are in concentric circles with the origin and insertion near the same point producing a sphincter to enclose an opening.

d. Pennate, in which the fascicles are short in relation to the muscle length and attach obliquely to the tendon, providing a very large number of fascicles distributed over the tendon, producing great power but a relatively smaller range of motion. (**Muscle architecture**)

Grade Yourself

Circle the numbers of the questions you missed, then fill in the total incorrect for each topic. If you answered more than three questions incorrectly, you need to focus on that topic. (If a topic has less than three questions and you had at least one wrong, we suggest you study that topic also. Read your textbook, a review book, or ask your teacher for help.)

Subject: Muscular System

Topic	Question Numbers	Number Incorrect
Types of muscle tissue	1, 2	
Characteristics of muscle tissue	3	
Anatomy of muscle tissue	4, 5, 6, 7, 8, 9, 10, 11, 12, 13, 14, 15, 16, 17	
Microscopic anatomy of muscle fibers	18, 19	
Contraction of muscles	20, 21, 22, 23, 24, 25, 26	
Muscle tension production	27	
Frequency of stimulation	28	
Length-tension relationship	29	
Motor unit recruitment	30	
Types of muscle contractions	31	
Muscle metabolism	32	
Muscle fatigue	33	
Muscle fiber types	34	
Cardiac muscle	35	
Smooth muscle	36, 37	
Skeletal muscle movements	38, 40	
Leverage	39	
Muscle architecture	41	

Nervous Tissue and the Peripheral Nervous System

Brief Yourself

One critical element to maintaining homeostasis is the ability of an organism to communicate the changes in the internal and external environments and to control the effects of any perturbations. The two systems of the body that share this responsibility are the nervous system and the endocrine system, each greatly influencing the effects of the other.

The nervous system differs from the endocrine system primarily with respect to the speed and duration of effects. The nervous system produces very rapid and transient effects, suited to respond to rapid changes in the environment, while the endocrine system typically produces effects that develop more slowly and tend to have a prolonged effect, suited to long term environmental alterations. The nervous system has three basic functions; sensory, integrative, and motor.

The nervous system can be broken into two primary branches, the central nervous system (CNS) and peripheral nervous system (PNS). The peripheral nervous system consists of the cranial and spinal nerves and their resulting branches.

Test Yourself

1. What are the sub-divisions of the PNS and their functions?

Questions 2–7 are matching. Match the following neuroglial cell with its description.

2. Found in the CNS, primarily responsible for the production of myelin for CNS axons — a. Astrocyte
3. Found in the CNS, produce cerebrospinal fluid — b. Oligodendrocyte
4. Found in the PNS, produce myelin around a single PNS axon — c. Microglia

5.	Found in the CNS, form a link between blood vessels and neurons	d.	Ependymal
6.	Found in the PNS, support neuron somas in ganglion	e.	Schwann (neurolemmocyte)
7.	Found in the CNS, phagocytic cells that destroy microbes and cellular debris	f.	Satellite

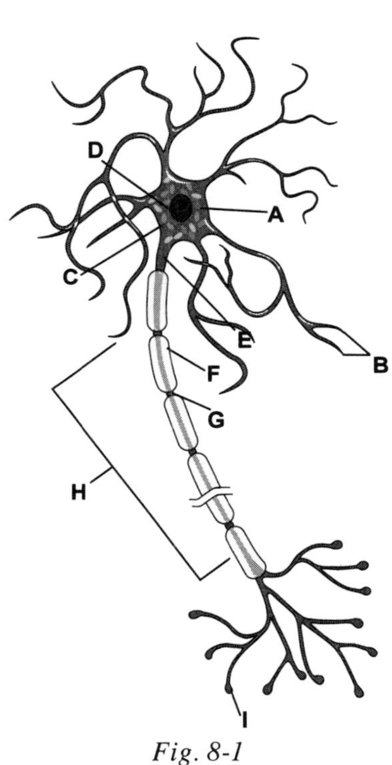

Fig. 8-1

8. Label the diagram of the neuron found below.

9. What is axoplasmic transport? Why is it important?

10. How are neurons classified on the basis of structure? On the basis of function?

11. What is an excitable membrane? What is meant by a selectively permeable membrane?

12. What is a resting membrane potential? What is the role of membrane ion pumps?

13. What are ion channels?

14. What is a graded potential and where do they occur on neurons?

15. What are action potentias and where do they occur on neurons?

16. What is primarily responsible for the propagation of an action potential and what is the role of the myelin sheath?

17. What is a refractory period and what causes it?

18. What can cause variations in the speed of conduction of action potentials?

19. How are impulses transmitted at the synapse? What produces excitatory post-synaptic potentials versus inhibitory post-synaptic potentials?

20. What are spatial and temporal summation of post-synaptic potentials?

21. What are the five basic chemical classifications of neurotransmitters?

22. What are four basic types of neuronal circuits?

23. Label the diagram of a typical spinal nerve.

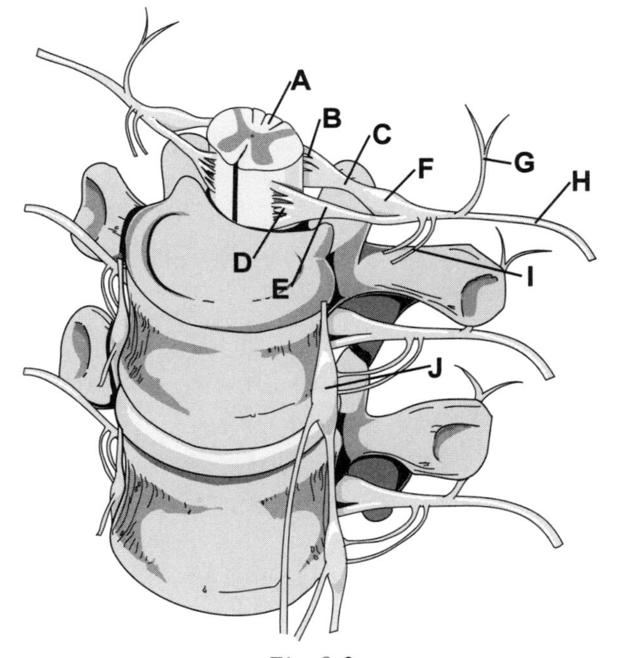

Fig. 8-2

Questions 24–30 are matching. Match the following term with its definition.

24. Branches of a spinal nerve following its exit from the intervertebral foramen
25. Fascia surrounding an individual axon in a spinal or cranial nerve
26. A network of joined anterior rami from several spinal nerves
27. Fascia surround a group of axons (fascicle) in a spinal or cranial nerve
28. The area of the skin that provides sensory input to one pair of spinal or cranial nerves
29. Fascia that surrounds an entire spinal or cranial nerve
30. Muscles innervated by the motor neurons of a single spinal segment

a. Endoneurium
b. Perineurium
c. Epineurium
d. Ramus
e. Plexus
f. Dermatome
g. Myotome

31. Define a spinal reflex arc and give its function. What are its five components?

32. What is the difference between ipsilateral, contralateral, and intersegmental reflexes?

Questions 33–44 are matching. Match the following cranial nerve with its location and function.

33. Arises from inner ear, purely sensory for audition and equilibrium
34. Arises from pons, largest cranial nerve, sensory from face, tongue, mouth, and motor to some muscles of the mandible
35. Arises at the receptors of the nasal cavity, purely sensory for olfaction
36. Arises from the dorsal surface of the midbrain, motor to superior oblique eye muscle and sensory from the proprioceptors of that muscle
37. Arises from a union of a cranial and spinal root, motor to the muscles of the larynx, pharynx, and neck and sensory from the proprioceptors of the same muscles
38. Arises from the retina of the eye, purely sensory for vision

a. Olfactory (I)
b. Optic (II)
c. Oculomotor (III)
d. Trochlear (IV)
e. Trigeminal (V)
f. Abducens (VI)

39. Arises from the medulla, motor to the tongue and pharynx and sensory from the tongue (taste), pharynx and carotid body — g. Facial (VII)

40. Arises from medulla and is only cranial nerve to extend below the area of the head and neck, motor and sensory to the parasympathetic nervous system — h. Vestibulocochlear (VIII)

41. Arises from the pons, motor to the lateral rectus muscle of the eye and sensory from the proprioceptors of the same muscle — i. Glossopharyngeal (IX)

42. Arises from the ventral midbrain, motor to both extrinsic and intrinsic muscles of eye with sensory from the proprioceptors of the same muscles — j. Vagus (X)

43. Arises from a series of medullary roots, motor to the muscles of the tongue and sensory from the proprioceptors of the same muscles — k. Accessory (XI)

44. Arises from lateral pons, motor to the muscles of facial expression, parasympathetic to lacrimal, nasal, and palatine glands and sensory from tongue (taste) — l. Hypoglossal (XII)

Check Yourself

1. Neurons conducting impulses from the PNS to the CNS are called sensory (afferent) while the neurons conducting impulses from the CNS to the PNS are called motor (efferent). The PNS is divided into a somatic and autonomic division, with somatic consisting of neurons conducting impulses to and from cutaneous and special receptors, and the skeletal muscular system. The autonomic system (ANS) consists of neurons conducting impulses to visceral organs, and may be further divided into the sympathetic and parasympathetic divisions. **(Divisions of the peripheral nervous system)**

2. b **(Neuroglia)**

3. d **(Neuroglia)**

4. e **(Neuroglia)**

5. a **(Neuroglia)**

6. f **(Neuroglia)**

7. c **(Neuroglia)**

8. a. soma

   b. dendrites

   c. Nissl bodies

   d. nucleus

   e. Axon Hillock

   f. Schwann cell (Myelin)

   g. Node of ranvier

   h. axon

   i. Axon terminal (Synaptic end bulb) **(Neuron anatomy)**

9. Most of the synthesis in the physiological processes of the neuron occur in the soma, which is often at some distance from the axon terminals. In order to move the synthesized chemicals (such as neurotransmitters or their precursors) to the site of their use, particles must be moved down the axons. Recycling of such chemicals after their use (recovered from the synaptic cleft by active transport) also requires their shipment up the axon to the soma for reuse. Both a slow and fast axoplasmic transport system exits, slow due to normal sarcoplasmic flow, and the fast system utilizing protein carriers and a microtubule system to move specific particles. **(Axoplasmic transport)**

10. Neurons are classified by structure as:

    a. multipolar, with many dendrites extending from the soma and one axon.

    b. bipolar, with one main dendrite extending from the soma and one axon.

    c. unipolar, with one main process extending from the soma so that the soma appears to be "set off to the side." Both branches of the process have the characteristics of an axon. Neurons are also classified by function as sensory neurons if they carry impulses from the PNS to the CNS, motor neurons if they carry impulses from the CNS to the PNS, or interneurons (association) if they are not specifically motor or sensory in function. **(Classification of neurons)**

11. An electrically excitable membrane is a membrane which will conduct electric events. Membranes must carry a resting electric charge (resting membrane potential) in order for this phenomenon to occur. The existence of a selectively permeable membrane, specifically in its permeability to ions, will provide the electrical nature necessary for such events. **(Neurophysiology—membrane potentials)**

12. In neurons, the resting membrane potential is about $^{-}70$mv, with the intracellular fluid approximately 70 mv more negative than the extracellular fluid when the membrane is at rest. This electrical potential exists due to the differences in concentrations of the cations and anions on either side of the membrane. Extracellular fluid is rich in Na^+ and Cl^-, while the intracellular fluid is rich in K^+ and organic anions of amino acids (in proteins) and phosphates. In particular, the distribution of Na^+ becomes a critical ion in determining the potential. The plasma membrane of neurons contain many K^+/Na^+ pumps which actively move Na^+ outside of the cell while actively moving K^+ inside. At rest, the membrane allows a certain permeability to K^+, but is practically impermeable to Na^+. The total distribution of ions results in an imbalance of charges that can be measured as the resting membrane potential. **(Neurophysiology—resting membrane potential)**

13. Ion channels allow for the diffusion of ions across the membrane and may either be "leaky" or "gated." Leaky channels are always open, but gated channels open and close in response to stimuli. Such channels are typically specific in the modality of the stimulus to which they respond. Four types of gated channels are known; voltage, chemical, mechanical, and light. Voltage channels open in response to changes in membrane potentials; chemical channels open in response to the binding of specific chemicals to receptors at the channel; mechanical channels open in response to vibration or pressure changes; light channels open in response to particular wavelengths of light. The ability of a membrane to open/close channels allows for the propagation of electrical impulses, as dramatic changes in the membrane potentials will occur when ions are allowed to diffuse across the membrane. **(Neurophysiology—ion channels)**

14. A graded potential is an alteration in the resting membrane potential due to the influx or efflux of ions across the membrane. This "flow" of ions occurs as ion channels are opened, allowing the ions to follow their electrical or chemical gradients. The size of these potentials varies with regard to the number of channels opened. These are localized events due to the normal loss of energy produced by resistance to electrical flow. A membrane potential change that produces less negativity across the membrane is called a depolarization and is typically produced by opening a Na^+ channel, which allows for the flow of Na^+ into the cell. A membrane potential change that produces more negativity across the membrane is called a hyperpolarization and is typically produced by opening Cl^- channels, which allows the flow of Cl^- into the cell or by opening additional K^+ channels which allows the flow of K^+ out of the cell. Typically, graded potentials occur primarily on the soma and dendrites of neurons. **(Neurophysiology—graded potentials)**

15. An action potential is a sequence of rapidly occuring events that decrease and eventually reverse the resting membrane potential and then restore it. Found on axons, these impulses can travel great distances due to the nature of the axon and as such are utilized for primary communication between neurons and their target effectors. **(Neurophysiology—action potentials)**

16. Voltage gates are primarily responsible for the propagation of the action potential. The membrane of the axon, beginning at its connection to the soma (the axon hillock), has a critical difference from the membrane of the soma or dendrites. The membrane of the axon contains many voltage-gated Na^+ channels which have a threshold to open of about -55mv. If electrical events (graded potentials) that are occurring on the dendrites or soma result in lowering the resting membrane potential to -55mv in the area of the axon hillock, these voltage channels for Na^+ open, producing a rapid depolarization in that area as Na^+ rushes into the cell. The resulting electrical event, called a spike, will often produce a local membrane potential of +40mv due to the Na^+ influx. This event will trigger the opening of the next voltage-gated Na^+ channel, producing a new influx, and so on, producing an electrical event that "moves" down the axon without a loss of energy. Myelinated axons have some distinct advantages. The myelin sheath acts as an insulator, allowing the current to flow through the cytosol without a loss of energy across the membrane. Only at the nodes of Ranvier does the ion current flow across the membrane (due to the opening of voltage-gated Na^+ channels at this point). The impulse appears to "jump" from node to node, producing a much faster conduction velocity. It is also more efficient in that Na^+ flow only occurs at the nodes and thus less energy is required to reset the resting membrane potential following an action potential. **(Neurophysiology—action potentials)**

17. The refractory period is the time it takes to reset the resting membrane potential following an action potential so that the axon can then propagate a second action potential. It is the time necessary to reestablish the original distribution of ions, primarily accomplished by activating the Na^+/K^+ pumps to remove the intracellular Na^+ that entered during the action potential. ATP is required to reset the resting membrane potential. **(Neurophysiology—refractory period)**

18. Action potentials are either propagated by continuous conduction or by saltatory conduction. Saltatory conduction utilizes graded potentials flowing beneath the myelin sheath to produce a more rapid and efficient conduction, producing the ionic current flowing across the membrane only at the nodes of Ranvier. In addition, the larger the diameter of the axon, the less resistance to electrical flow and the more rapid the conduction becomes. **(Neurophysiology—conduction velocity)**

19. At the synapese between an axon and its effector cell, impulses are transmitted either electrically or chemically. Electrical synapses occur where the cell membranes of the communicating cells are contiguous with one another at special membrane structures called gap junctions which are composed of connexons. These tubular structures allow the flow of cytosol between the cells and thus allows the transmission of ionic currents directly. Chemical synapses utilize a neurotransmitter released by the pre-synaptic membrane as a result of the arrival of action potentials which diffuses across the synaptic cleft. There it binds with receptors on the post-synaptic membrane and chemically-gated ion channels can be opened. If the channels affected produce a depolarization due to the influx of Na^+, the event on the effector membrane is called an excitatory post-synaptic potential (EPSP). If the channels affects produce a hyperpolarization due to the influx of Cl^- or efflux of K^+, the event is an inhibitory post-synaptic potential. **(Synaptic transmission)**

20. Post-synaptic potentials only last a few milliseconds, but a typical neuron will receive inputs from many sources that will be rapidly applied. The integration of the resulting electrical events occurs at the spiking zone of the axon hillock and is called summation. Spatial summation occurs when there is a buildup of neurotransmitter (and thus post-synaptic potentials) released on a post-synaptic membrane from a variety of locations (from many synapses). Temporal summation occurs when there is a buildup of neurotransmitter released on a post-synaptic membrane due to repeated releases from a single synapse. **(Summation)**

21. Neurotransmitters are found in five chemical groups:

    a. Cholinergic, such as acetylcholine.

    b. Biogenic amines, such as norepinephrine or dopamine.

    c. Amino acids, such as gamma-aminobutyric acid.

    d. Peptides, such as endorphins.

    e. Novel, such as nitrous oxide or carbon monoxide. **(Neurotransmitters)**

22. Neuronal groups, especially in the CNS, are organized into patterns called neural circuits. The four basic neural circuits are:

    a. Diverging, such that a single neuron, through the branching of its axon, can influence a number of other neurons.

    b. Converging, such that a number of neurons will influence a neuron.

    c. Reverberating, such that a neuron influences a chain of neurons, each with collaterals that synapse with previous neurons in the chain, allowing for impulses moving through the chain to be repeated by feedback loops.

    d. Parallel, such that a single neuron will influence a group of neurons, all of which will influence a single neuron. **(Neural circuits)**

23. a. spinal cord

    b. posterior rootlets

    c. posterior root

    d. anterior rootlets

    e. anterior root

    f. posterior root ganglion

    g. dorsal ramus

    h. ventral ramus

    i. rami communicantes

    j. sympathetic ganglion **(Anatomy of spinal nerve)**

24. d **(Spinal nerves)**

25. a **(Spinal nerves)**

26. e **(Spinal nerves)**

27. b **(Spinal nerves)**

28. f **(Spinal nerves)**

29. c **(Spinal nerves)**

30. g **(Spinal nerves)**

31. A reflex is a predictable, automatic response to a particular environmental stimulus. Some reflexes are integrated in the brain and involve cranial nerves (cranial reflexes), but many are integrated in the gray matter of the spinal cord and are called spinal reflexes. Spinal reflexes provide for very rapid responses to stimuli and do not require higher processing in the brain. Each spinal reflex has five components:

    a. Receptor, the distal end of a sensory neuron that associates with a specialized structure, often a specialized epithelial cell that will produce generator (receptor) potentials in that sensory neuron. If strong enough, the generator potentials will cause a depolarization and action potential in the neuron.

    b. Sensory neuron, it propogates the impulses to the spinal cord gray matter.

    c. Integration center, one or more association (neurons) interneurons that act to modify or relay the impulse to motor neurons.

    d. Motor neuron, propagates the impulse to the effector tissue.

    e. Effector, the responding organ which may be a gland or muscle. If the impulse goes to a skeletal muscle, the reflex is a somatic reflex. If the impulse goes to a gland or other visceral organ, it is an autonomic reflex. **(Spinal reflexes)**

32. Ipsilateral spinal reflexes are those in which the sensory nerve impulses enter the spinal cord on the same side from which the motor responses will be issued. All monosynaptic reflex arcs are ipsilateral. Contralateral spinal reflexes are those in which sensory nerve impulses enter the spinal cord from the opposite side from which the motor responses will be issued. These typically indicate a higher degree of processing than the monosynaptic reflex arc. Intersegmental reflexes are those in which the sensory nerve impulses entering the spinal cord are distributed to several motor neuron pools along the spinal cord allowing for the control of a group of muscles in one reflex. **(Spinal reflexes)**

33. h **(Cranial nerves)**

34. e **(Cranial nerves)**

35. a **(Cranial nerves)**

36. d **(Cranial nerves)**

37. k **(Cranial nerves)**

38. b **(Cranial nerves)**

39. i **(Cranial nerves)**

40. j **(Cranial nerves)**

41. f **(Cranial nerves)**

42. c **(Cranial nerves)**

43. l **(Cranial nerves)**

44. g **(Cranial nerves)**

Grade Yourself

Circle the numbers of the questions you missed, then fill in the total incorrect for each topic. If you answered more than three questions incorrectly, you need to focus on that topic. (If a topic has less than three questions and you had at least one wrong, we suggest you study that topic also. Read your textbook, a review book, or ask your teacher for help.)

Subject: Nervous Tissue and the Peripheral Nervous System

Topic	Question Numbers	Number Incorrect
Divisions of the peripheral nervous system	1	
Neuroglia	2, 3, 4, 5, 6, 7	
Neuron anatomy	8	
Axoplasmic transport	9	
Classification of neurons	10	
Neurophysiology—membrane potentials	11	
Neurophysiology—resting membrane potential	12	
Neurophysiology—ion channels	13	
Neurophysiology—graded potentials	14	
Neurophysiology—action potentials	15, 16	
Neurophysiology—refractory period	17	
Neurophysiology—conduction velocity	18	
Synaptic transmission	19	
Summation	20	
Neurotransmitters	21	
Neural circuits	22	
Anatomy of spinal nerve	23	
Spinal nerves	24, 25, 26, 27, 28, 29, 30	
Spinal reflexes	31, 32	
Cranial nerves	33, 34, 35, 36, 37, 38, 39, 40, 41, 42, 43, 44	

The Central Nervous System, General Sensory, and Motor Processing

Brief Yourself

The central nervous system (CNS) consists of the brain and spinal cord. While the peripheral nervous system acts to gather information from the body and to distribute response to it, the CNS is an amazing integrator of the information gathered and is responsible for the design of appropriate responses that vary in complexity from the simple monosynaptic reflex to the impressive ability of conscious thought.

As a part of the course of the evolutionary development of the nervous system, a great degree of elaboration and centralization has occurred near the rostral or anterior portion of the nervous system called cephalization. This cephalization is its most impressive in the production of the brain, an incredibly complex and flexible processor of information. More than any other organ of the body, the brain is responsible for the adaptability of man to many environments.

Test Yourself

1. Describe the position of the spinal cord in the body and the bony covering of the spinal cord.

2. Describe the spinal meninges.

3. Label the following diagram of a the major components of the nervous system.

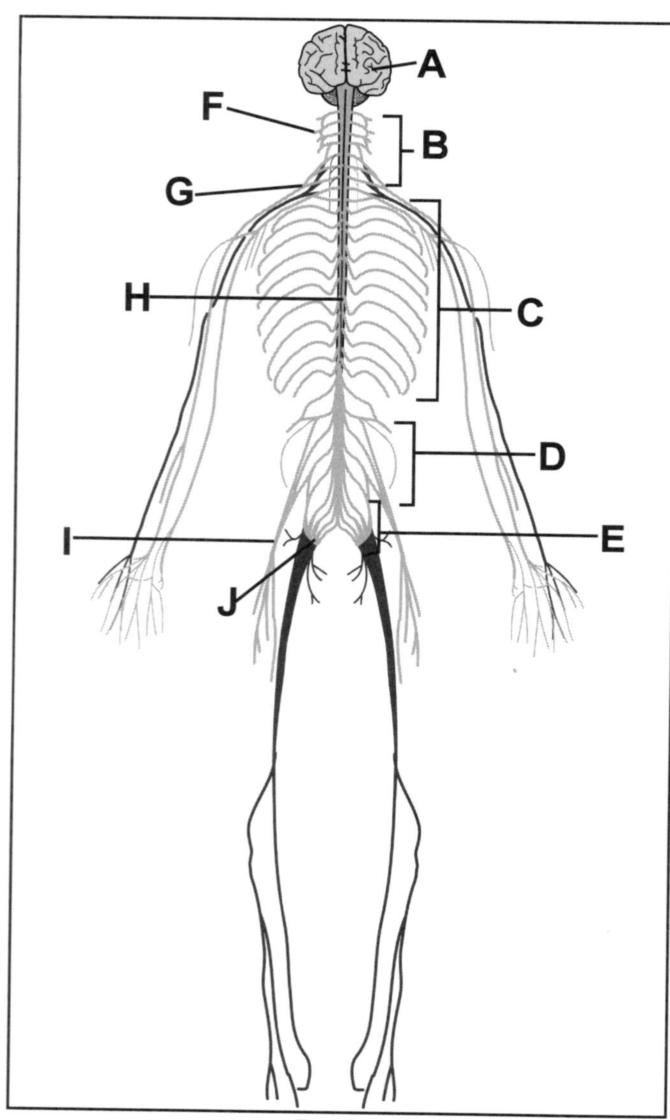

Fig. 9-1

4. Label the following diagram of a cross-section of the spinal cord.

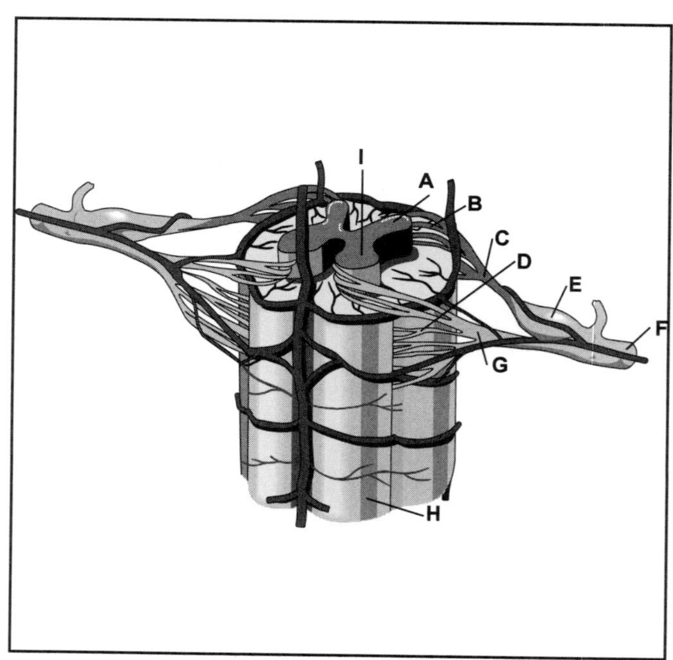

Fig. 9-2

Questions 5–13 are matching. Match the spinal pathway with its description.

5. Conveys sensory information for pain, thermal, and some tactile information to the thalamus on the opposite side of the body
6. Conveys motor information from the superior colliculus to skeletal muscles of the neck, head, and eye on the opposite side of the body for visual tracking
7. Conveys sensory information for conscious proprioception and tactile information from one side of the body to the medulla oblongata of the opposite side
8. Conveys motor information from the vestibular nucleus to the muscles of the same side of the body for maintenance of balance
9. Conveys motor information from the pontine and medullary reticular formations to the muscles of the axial skeleton and proximal appendages for maintenance of muscle tone and mediation of spinal somatic reflexes
10. Conveys motor information from the motor cortex to the muscles of the opposite side of the body to initate and coordinate conscious movements
11. Conveys sensory information for subconscious proprioception to the cerebellum
12. Conveys motor information from the red nucleus to the muscles of the opposite side of the body to coordinate fine movements by the hands and feet
13. Conveys motor information from the motor cortex to the muscles that initiate and coordinate fine conscious movements of the head and neck

a. Posterior columns (fasciculus gracilis and cuneatus)
b. Spino-cerebellar tracts
c. Spino-thalamic tracts
d. Cortico-spinal tracts
e. Reticulo-spinal tracts
f. Rubrospinal tracts
g. Olivospinal tracts
h. Vesti-bulospinal tracts
i. Tectospinal tracts

14. Identify the primary vesicles of the brain. What are the secondary vesicles? What principal components of the brain are contained in each?

15. Describe the cranial meninges.

16. What is cerebrospinal fluid (CSF) and what is its function? How and where is CSF made?

17. What are the ventricles of the brain? List them and give their function.

18. Label this diagram of a mid-sagittal section of the brain.

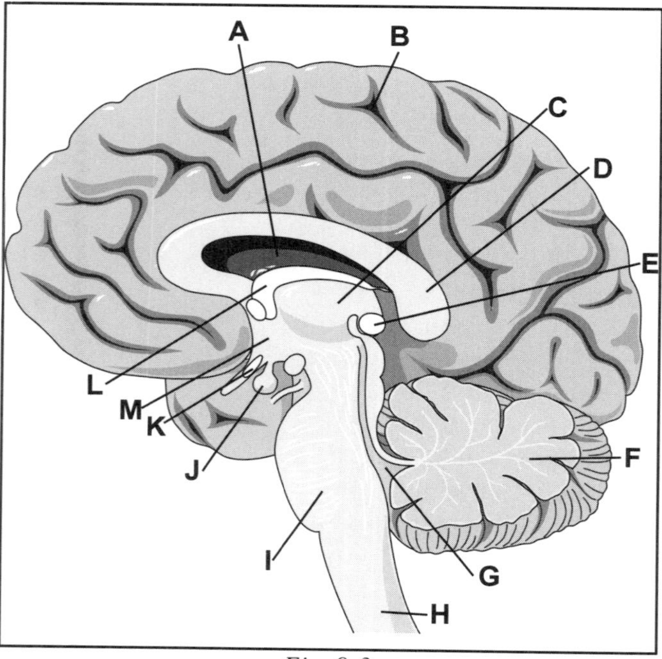

Fig. 9-3

19. Describe the structure and function of the medulla oblongata.

20. Describe the structure and function of the pons.

21. Describe the structure and function of the midbrain.

22. Describe the structure and function of the cerebellum.

23. Describe the structure and function of the thalamus.

24. Describe the structure and function of the hypothalamus.

25. Label the following diagram of the lobes and domains (functional areas) of the cerebrum.

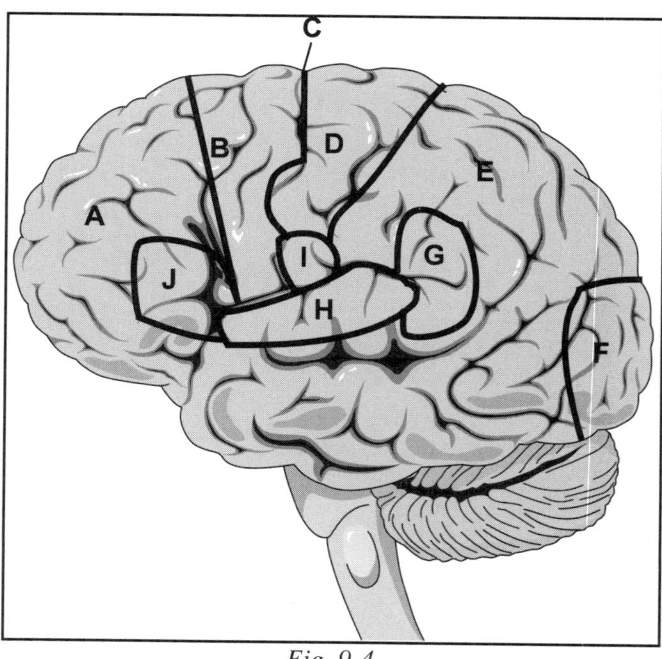

Fig. 9-4

26. Identify and give the function of the white fiber tracts found in the cerebrum.

27. List the nuclei of the basal ganglia and give the function of each.

28. What is the limbic system? What are its components and functions?

29. What is the primary motor cortex? What is its function?

30. What is the primary somatosensory cortex? What is its function?

31. What are association areas of the cortex? What are their function?

32. What is the difference between sensation and perception?

33. What is meant by modality of sensation? What are the components of sensation?

34. How might sensory receptors be classified according to location? According to modality?

35. List and describe the cutaneous receptors.

36. What is pain? What types of pain exist and what receptors are responsible for it?

37. What are proprioceptive receptors? List the three major types.

38. Describe the structure and function of a muscle spindle. How can it continue to provide for length information as a muscle is actively changing its length?

39. What is sensory-motor integration? Why is it important?

40. What are the direct (pyramidal) motor pathways and their function?

41. What are the indirect (extrapyramidal) motor pathways and their function?

42. How might the primary motor cortex, basal ganglia, and cerebellum all function together in the production and coordination of ongoing movements?

43. What are learning and memory? What is memory consolidation and how is it thought to occur?

44. What are wakefulness and sleep? What serves to control these states?

Check Yourself

1. The spinal cord is found on the dorsal (posterior) aspect of the body, encased in the vertebral column. The vertebral foramina of all the vertebrae form the vertebral canal. **(Anatomy of the spinal cord)**

2. The spinal meninges are the connective tissue coverings that surround the spinal cord. The most superficial layer is the dura mater, a tough, fibrous connective tissue sheath. The middle layer is the arachnoid, with a spider's web network of collagen and elastic fibers that connect to the deepest layer, the pia mater. Between the dura mater and the arachnoid mater is the subdural space, filled with interstitial fluid. Between the arachnoid mater and the pia mater is the subarachnoid space, filled with cerebrospinal fluid. The pia mater adheres directly to the spinal cord and is composed of fine collagen and elastic fibers. **(Spinal meninges)**

3. a. Brain

   b. Cervical spinal nerves

   c. Thoracic spinal nerves

   d. Lumbar spinal nerves

   e. Sacral spinal nerves

   f. Cervical plexus

   g. Brachal plexus

   h. Spinal cord

   i. Lumbar plexus

   j. Sacral plexus **(Anatomy of the spinal cord)**

4. a. Posterior horn of gray matter

   b. Posterior rootlets

   c. Posterior root

   d. Anterior rootlets

   e. Posterior root ganglion

   f. Spinal nerve

   g. Postieror root

   h. White matter

i. Anterior horn **(Anatomy of the spinal cord)**

5. c **(Anatomy of the spinal cord)**

6. i **(Anatomy of the spinal cord)**

7. a **(Anatomy of the spinal cord)**

8. h **(Anatomy of the spinal cord)**

9. e **(Anatomy of the spinal cord)**

10. d **(Anatomy of the spinal cord)**

11. b **(Anatomy of the spinal cord)**

12. f **(Anatomy of the spinal cord)**

13. g **(Anatomy of the spinal cord)**

14. The primary vesicles of the brain are the rhombencephalon, mesencephalon, and prosencephalon. The rhombencephalon is further divided into the metencephalon and myelencephalon. The prosencephalon is further divided into the telencephalon and diencephalon. The major structures found in each are:

    a. myelencephalon; the medulla oblongata of the brainstem.

    b. metencephalon; the cerebellum and pons of the brainstem.

    c. mesencephalon; midbrain of the brainstem.

    d. diencephalon; the thalamus, hypothalamus, and epithalamus.

    e. telencephalon; the cerebrum and basal ganglia. **(Basic anatomy of the brain)**

15. The cranial meninges are continuous with the spinal meninges and have the same structure and layers, including the dura mater, arachnoid mater, and pia mater. The dura mater surrounding the brain has three predominate extensions. The falx cerebri invaginates to separate the two hemispheres of the cerebrum, the falx cerebelli invaginates to separate the two hemispheres of the cerebellum, and the tentorium cerebelli invaginates to separate the cerebellum from the cerebrum. **(Cranial meninges)**

16. Cerebrospinal fluid (CSF) is a special filtrate of the blood produced by the ependymal cells of the choroid plexuses of the brain. The choroid plexuses consist of capillaries covered with ependymal cells which are interconnected by tight junctions so that no fluid flow is available between the cells. This produces a blood-brain barrier such that substances released by the capillaries to the choroid plexuses must pass through the ependymal cells to reach the neural tissue. CSF serves as a shock absorber by filling the sub-arachnoid space, serving to "float" the neural tissue in the bony encasements, produces an optimal chemical environment for the electrical activity of neurons, and provides nutrients and waste removal. **(Cerebrospinal fluid)**

17. The ventricles of the brain are enlargements in the hollow canal inside the brain (formed from the neural tube). There are four ventricles: the two lateral ventricles supporting the hemispheres of the cerebrum, the third ventricle in the diencephalon, and the fourth ventricle located between the brainstem and the cerebellum. The major choroid plexuses of the brain are found in the ventricles. The connections between the ventricles, such as the interventricular foramen or cerebral aqueduct allows for the circulation of CSF. **(Ventricles of the brain)**

18. a. Lateral ventricle

    b. Cerebrum

    c. Thalamus

    d. Corpus callosum

    e. Pineal gland

    f. Cerebellum

    g. Fourth ventricle

    h. Spinal cord

    i. Pons

    j. Pituitary gland

    k. Optic nerve

    l. Fornix

    m. Hypothalamus **(Anatomy of meninges and ventricles)**

19. The medulla oblongata is the most inferior part of the brainstem and is continuous with the spinal cord. All of the ascending and descending tracts that connect the brain to the spinal cord pass through the medulla. It also contains a variety of nuclei that control body functions as well as nuclei that give rise to five of the cranial nerves. The major structures of the medulla include:

    a. medullary pyramids: two bulges on the anterior surface caused by the large motor fiber tracts that pass from the motor cortex of the cerebrum to the spinal cord.

    b. nucleus gracilis and nucleus cuneatus: on the dorsal aspect, these nuclei receive inputs from the fasciculi gracilis and cuneatus respectively. Second order neurons arise here to relay information to the thalamus.

    c. cardiovascular center: regulates heartrate and bloodflow.

    d. medullary rhythmicity area: adjust breathing rate and depth.

    e. nuclei associated with swallowing, coughing, sneezing, and vomiting.

f. nuclei of the vestibulocochlear, glossopharyngeal, vagus, accessory, and hypoglossal cranial nerves.

g. inferior olives: nuclei on the lateral aspect that communicate with the cerebellum through the inferior cerebellar peduncles.

h. the fourth ventricle: found just below the cerebellum. **(Medulla oblongata)**

20. The pons is found just superior to the medulla and appears as a bulge on the brainstem. It serves as a bridge for the connections of the brain to the spinal cord and to different parts of the brain. Its transverse fibers form the middle cerebellar peduncles, connecting the right and left sides of the cerebellum. Its longitudinal fibers connect the medulla to the superior areas of the brain. It also includes the pontine pneumotaxic and apneustic nuclei that work with the medullary rhythmicity center to set breathing rates. Nuclei for the trigeminal, abducens, facial, and some of the vestibular branches of the vestibulocochlear cranial nerves are found in the pons. **(Pons)**

21. The midbrain is found just superior to the pons and just inferior to the diencephalon. The cerebral aqueduct passes through the midbrain connecting the third ventricle above to the fourth ventricle below. It contains both tracts and regulatory nuclei. Its major components include:

    a. cerebral peduncles: tracts containing motor fibers descending from the motor cortex and other areas and sensory fibers arising from the medulla to the thalamus.

    b. superior cerebellar peduncles: tracts that connect the midbrain to the cerebellum

    c. superior colliculi: nuclei found in the tectum (posterior portion of the midbrain) that receive visual information and produce reflex responses mediated through the fibers that become the tectospinal motor tract in the spinal cord.

    d. inferior colliculi: nuclei found in the tectum, just inferior to the superior colliculi, that receive auditory information and produce reflex responses mediated through the fibers that become the tectospinal motor tract in the spinal cord.

    e. substantia nigra: nuclei containing darkly pigmented nuclei involved with the control of subconscious muscle activity.

    f. red nuclei: nuclei richly supplied with blood (and thus the iron pigment for color) which function through connections to the cerebral cortex, cerebellum, and basal ganglia to coordinate movement.

    g. nuclei of the occulomotor and trochlear cranial nerves.

    h. medial lemniscus: a sensory fiber tract arising from the medulla and pons that connects to the thalamus for conscious tactile and proprioceptive sensation. **(Midbrain)**

22. The cerebellum is found on the posterior aspect of the brainstem at the level of the pons. It is the second largest area of the brain, supported by the fourth ventricle and consists of two cerebellar hemispheres connected by a constricted area called the vermis. Each hemisphere is divided into three lobes; the anterior and posterior (middle) lobes which are concerned with subconscious control of skeletal muscles and the flocculonodular lobe concerned with equilibrium. It is attached to the brainstem by the superior, inferior, and middle cerebellar peduncles.

    The cerebellum typically functions subconsciously to compare motor programs produced in the motor cortex and other motor areas of the brain with the actual results of those programs by receiving proprioceptive information from the body. Feedback allows for error detection by the cerebellum which stimulate or inhibit motor areas of the brain to provide motor corrections to smooth and coordinate movements. **(Cerebellum)**

23. The thalamus is located in the diencephalon and consists of a cluster of nuclei and fiber tracts in two lobes connected by a bridge of gray matter called the intermediate mass. The thalamus is the major relay station for sensory information that ascends to the cerebrum. As well as serving as a major filter for sensory information rising to the cerebrum, the thalamus is responsible for some crude perception of non-localized sensation and plays a role in cognition in cooperation with the cerebrum. The thalamus contains several well delineated nuclei for specific modalities, including:

    a. medial geniculate: audition.

    b. lateral geniculate: vision.

    c. ventral posterior: taste, tactile and proprioception.

    d. ventral lateral: voluntary motor.

    e. ventral anterior: voluntary motor.

    f. anterior: emotion and memory. **(Thalamus)**

24. The hypothalamus is located in the diencephalon just inferior to the thalamus. It is divided into four major areas consisting of about a dozen nuclei. The hypothalamus is the major regulator of homeostasis of the body, receiving sensory information from virtually all the somatic and visceral organs as well as from most of the special senses. It also contains receptors which monitor the osmotic pressures in the body, blood temperature, and hormone concentrations. The regions of the hypothalamus include:

    a. mammillary region: adjacent to the midbrain, contains posterior hypothalamic nuclei and mammillary bodies which control reflex responses to the olfactory information they receive.

    b. tuberal region: contains the dorsomedial, ventromedial, and arcuate nuclei as well as giving rise to the infundibulum for connections to the pituitary gland.

    c. supraoptic region: superior to the optic chiasm, contains paraventricular, supraoptic, anterior hypothalamic, and suprachiasmatic nuclei.

    d. preoptic region: preoptic periventricular, medial periventricular, medial preoptic, and lateral preoptic nuclei. The functions of the hypothalamus include:

        (1) control of the ANS; fiber tracts arise in the hypothalamus that descend to synapse with the sympathetic and parasympathetic ganglia of the brainstem and spinal cord.

        (2) control of the pituitary gland; neurons of the hypothalamus produce hypothalamic regulating hormones that are released into the pituitary to stimulate or inhibit the production of pituitary hormones also neurosecretory cells of the hypothalamus produce and deliver antidiuretic hormone and oxytocin to the pituitary where it is stored for release.

        (3) regulation of emotion; works with the limbic system to regulate emotional behavior

(4) regulation of eating and drinking; has a feeding center reponsible for hunger sensations and a satiety center as well as a thirst center which responds to the osmotic pressure receptors of the hypothalamus.

(5) control of body temperature; directs responses of the ANS with regard to blood temperature.

(6) regulation of diurnal rhythms; in cooperation with other areas of the brain, regulates patterns of sleep. **(Hypothalamus)**

25. a. Frontal cortex

    b. Primary motor area

    c. Central sulcus

    d. Primary somatosensory area

    e. Parietal cortex

    f. Primary visual area

    g. Wernicke's area

    h. Primary auditory area

    i. Primary gustatory area

    j. Broca's area **(Functional areas of the cerebrum)**

26. The fiber tracts of the cerebrum are classified as:

    a. association fibers; transmit impulses between the gyri of the same hemisphere.

    b. commissural fibers; transmit impulses from the gyri of one hemisphere to the corresponding gyri of the other hemisphere.

    c. projection fibers; transmit impulses from the cerebrum to lower centers of the brain or spinal cord and transmit impulses from the lower centers or spinal cord to the cerebrum. **(Fiber tracts of the cerebrum)**

27. The basal ganglia are a group of nuclei found in each cerebral hemisphere inferior to the cerebral cortex. These nuclei receive and process information from the cerebral cortex, thalamus, and hypothalamus. The output of the basal ganglia forms a part of the motor programming of the brain. The nuclei of the basal ganglia include the corpus striatum which is further divided into the caudate and lenticular nuclei. The lenticular nucleus is subdivided laterally into the putamen and medially into the globus pallidus. The substantia nigra and red nucleus of the brainstem are sometimes considered to be a part of the basal ganglia due to their functional links. **(Basal ganglia)**

28. The limbic system is a network of nuclei that stretch from the hypothalamus of the diencephalon to cerebrum. The limbic system is responsible for the control of emotional behaviors such as pain, pleasure,

rage, fear, sorrow, affection, sexual arousal, etc. It also functions in the production (consolidation) of memories. The components of the limbic system include:

a. parahippocampal and cingulate gyri of the cerebral cortex.

b. hippocampus: an extension of the parahippocampal gyrus.

c. dentate gyrus.

d. amygdala.

e. mammillary bodies of the hypothalamus.

f. anterior nucleus of the thalamus.

g. olfactory bulbs. **(Limbic system)**

29. The primary motor cortex is located on the precentral gyri of the frontal lobes. Its neurons are responsible for the initiation of voluntary motor activity and cooperate with the basal ganglia in the production of motor programs. **(Primary motor cortex)**

30. Primary somatosensory cortex is located on the post-central gyri of the parietal lobes. Its neurons receive somatic sensory information for tactile, proprioception, pain, and temperature and serve to localize the position of the stimuli with regard to their position on the body. This area cooperates with other areas of the cortex and thalamus in the production of somatosensory perceptions. **(Primary somatosensory cortex)**

31. Association areas of the cortex are non-specific areas of the cortex that are involved with higher integrative function between incoming sensation, emotional states, stored information (memories), and motor programming. The concept of conscious thought or intelligence is believed to be primarily a result of the activity of the association areas of the cortex. **(Association areas of the cortex)**

32. Sensation is the simple awareness, either conscious or subconscious, of external or internal stimuli that reach a threshold such that the neurons associated with the receptors of these stimuli produce and propagate impulses to the CNS. Perception is the integration and interpretation of the sensory input rising to the CNS to produce conscious cognition. **(Sensation and perception)**

33. The modality of sensation is the specific type of stimuli to which our receptors respond. For example, the visual receptors of the retina only respond to light within certain wavelengths and to no other forms of energy. For this reason, impulses sent from the retina to the cortex are perceived as light information. Sensations have the following components:

    a. stimulus; a change in the environment of sufficient nature to activate certain sensory neurons.

    b. transduction; the stimuli must be transduced (transformed) into a generator potential by the sensory receptor.

    c. impulse conduction; the generator potential must elicit a neural impulse that is propagated to the CNS along a first-order sensory neuron.

    d. integration; a region of the CNS must receive the impulse and integrate it to produce a sensation. **(Sensation and perception)**

34. Two classification systems exist for sensory receptors. By location, receptors are said to be exteroceptors when located on the surface of the body, interoceptors when located in the visceral organs, and proprioceptors when located in the musculoskeletal system. Receptors are also classified according to the modality of stimulus to which they will respond. Mechanoreceptor respond to mechanical energy, thermoreceptors respond to thermal energy, nociceptors respond to damage to the tissues (pain), photoreceptors respond to electromagnetic radiation (light), and chemoreceptors respond to chemical energy. **(Sensory receptors)**

35. The cutaneous (skin) receptors include:

    a. Meissener's corpuscles: located in the dermal papillae for discriminative touch.

    b. hair root plexi: located around hair follicles for sensation of hair movement.

    c. Merkel's discs: found in the stratum basale for discriminative touch.

    d. Ruffini endings: found deep in the dermis for heavy touch.

    e. Pacinian corpuscles: widely distributed in subcutaneous, visceral organs, joints, etc. for pressure sensations.

    f. free nerve endings: found in almost every tissue of the body, responsible for pain and temperature sensations. **(Cutaneous receptors)**

36. Pain is the perception of tissue-damaging stimuli. It is produced primarily as a result of the impulses carried by free nerve endings. Pain is classified as acute or chronic. Acute pain is produced by the response of myelinated free nerve ending and occurs rapidly after the stimulus is applied. It is described as a "sharp" pain found in the responses of the receptors of the skin. Chronic pain occurs more gradually and typically builds in intensity over time, carried by smaller, unmyelinated free nerve endings. This pain is considered a "dull" pain but can be excruciating. It arises from the deep tissues of the body as well as from the skin. **(Nociceptors and pain)**

37. Proprioceptive sensations provide an awareness of the position of the body in space. The three major proprioceptive receptors are the muscle spindles which provide information about the length of muscles, the Golgi tendon organs which provide information about the force applied to tendons, and the joint kinesthetic receptors which provide information about pressure, vibration, and position in the joint capsules. **(Proprioceptors)**

38. A muscle spindle is a unique sensory organ that contains both motor and sensory capabilities. It consists of a group of specialized muscle fibers and nerve endings that are interspersed among the skeletal muscle fibers in a muscle. The spindle fibers are parallel to the skeletal fibers and are anchored to the endomysium and perimysium of the muscle. The spindle has 3-10 intrafusal muscle fibers, each with their own innervation by gamma motorneurons.

    A central component of each intrafusal fiber is non-contractile and is innervated by two sensory nerve endings, type Ia and type II fibers. When the central portion of the intrafusal fiber is stretched, these endings are stimulated and they send impulses to the CNS. Since the intrafusal fibers lie in parallel with the extrafusal muscle fibers (normal skeletal muscle fibers), any stretch on a resting muscle will stretch the intrafusals and thus sensory information will be propagated to the CNS from the type Ia and type II endings about the length of the muscle. However, a contracting (shortening) muscle would serve to "unload" the intrafusals as the larger extrafusals shortened. In order to continue to serve as an effective sensory organ for muscle length, the intrafusals must be able to shorten at the same rate as the extrafusals.

This is possible due to the innervation of the gamma motor neurons to the contractile elements of the intrafusal fibers, allowing them to "keep pace" with the extrafusal fibers. **(Muscle spindles)**

39. Sensory-motor integration is the process of analyzing afferent information and producing appropriate efferent responses. The CNS is constantly receiving a great deal of sensory information concerning conditions in the external and internal environment. This information is assembled and integrated to produce a perception at the subconscious and conscious level about the state of the body in the external environment and the state of the internal environment. Structures of the cerebral cortex, basal ganglia, cerebellum, and brainstem cooperate in selecting and developing appropriate motor responses to the incoming sensations. **(Sensory-motor integration)**

40. The direct motor pathways are the fibers projecting from the neurons of the motor cortex or from motor nuclei of the cranial nerves to the motorneuron cell bodies of the anterior horn of the spinal cord. The direct pathways conduct impulses that result in precise voluntary movements. **(Direct motor pathways)**

41. The indirect motor pathways include all of the descending motor tracts other than the corticospinal and corticobulbar tracts. These motor impulses typically follow complex pathways that include circuits that connect the motor cortex, basal ganglia, cerebellum, limbic system, thalamus, reticular system, and other brainstem nuclei. This system is thought to be responsible for much of the subconscious motor programming and motor control, both for voluntary and involuntary movements. **(Indirect motor pathways)**

42. The motor programming and control of voluntary movement is very complex. It is believed that the conscious intent to move arises in the association areas of the cerebral cortex. Once such an intent arises, it becomes the task of the motor cortex and associated motor areas to select or design an appropriate motor program.

    It appears that such programs are produced by the cooperative efforts of the motor cortex and basal ganglia as mediated through the thalamus. Once the program is complete, the movement it probably initiated by the direct pathways from the motor cortex. At the same time, communication between the motor cortex, basal ganglia, and cerebellum occurs providing the cerebellum with information concerning the "desired" outcome of the movement. The cerebellum can then monitor proprioceptive feedback generated by the movement and can induce small corrections to smooth and coordinate the movement. If large corrections are required, the motor cortex and basal ganglia are "notified" and re-programming can be produced. **(Motor programming)**

43. Learning and memory are virtually inseparable phenomena. Learning is defined as a relatively permanent behavioral change due to instruction or experience. Memory is the process by which information is retained. Without memory of past instruction or experience, learning would be impossible. Memory is classified into two broad areas, short-term memory and long-term memory. Short-term memory is the ability to recall a few bits of information for a brief period (a few seconds). Long-term memory is the ability to recall information over much longer periods of time, often for a lifetime.

    The ability to take short-term memory and convert it to long-term memory is called consolidation. It is theorized that short-term memory, our "conscious processor," is produced by the production of neural activity, perhaps in the form of reverberating circuits. Your ability to recall the information only lasts as long as the circuit reverberates. Long-term memory is thought to occur due to structural changes in the circuits of the brain, consolidating these "new" circuits into stable ones. Thus, forgetting is the decay of stable circuits, and our new memories are the production of new, stable circuits. **(Learning and memory)**

44. Wakefulness and sleep are conditions of activation in the CNS. Wakefulness is thought of as consciousness and is characterized by great amounts of activity of the cerebral cortex in particular. Sleep

is a state of altered consciousness that is characterized by low activity of the cerebral cortex. These differences in the level of arousal are under the control of the reticular activating system. When active, this system transmits impulses through the thalamus to the cortex producing an "activation" or increase in cortical activity associated with wakefulness. The reticular system receives sensory information from almost all modalities and as such is alerted by most stimuli. Lowering the (sensory information) feedback entering the reticular system can help to induce sleep, such as finding a comfortable, dark, quiet area, etc. **(Sleep and arousal)**

Grade Yourself

Circle the numbers of the questions you missed, then fill in the total incorrect for each topic. If you answered more than three questions incorrectly, you need to focus on that topic. (If a topic has less than three questions and you had at least one wrong, we suggest you study that topic also. Read your textbook, a review book, or ask your teacher for help.)

Subject: The Central Nervous System General Sensory, and Motor Processing

Topic	Question Numbers	Number Incorrect
Anatomy of the spinal cord	1, 3, 4	
Spinal meninges	2	
Basic anatomy of the spinal cord	5, 6, 7, 8, 9, 10, 11, 12, 13, 14	
Cranial meninges	15	
Cerebrospinal fluid	16	
Ventricles of the brain	17	
Anatomy of meninges and ventricles	18	
Medulla oblongata	19	
Pons	20	
Midbrain	21	
Cerebellum	22	
Thalamus	23	
Hypothalamus	24	
Functional areas of the cerebrum	25	
Fiber tracts of the cerebrum	26	
Basal ganglia	27	
Limbic system	28	

Topic	Question Numbers	Number Incorrect
Primary motor cortex	29	
Primary somatosensory cortex	30	
Association areas of the cortex	31	
Sensation and perception	32, 33	
Sensory receptors	34	
Cutaneous receptors	35	
Nociceptors and pain	36	
Proprioceptors	37	
Muscle spindles	38	
Sensory-motor integration	39	
Direct motor pathways	40	
Indirect motor pathways	41	
Motor programming	42	
Learning and memory	43	
Sleep and arousal	44	

The Special Senses

10

Brief Yourself

It is often said that humans have five senses; touch, smell, taste, sight, and hearing. Our sense of touch is produced by the complex group of general senses that do include tactile information, but are also composed of proprioceptive, thermal, pain, and mechanical information from the skin and the musculoskeletal system, as well as from other visceral organs. In contrast, our special senses are produced by special sensory receptor cells that are found organized into special sensory organs.

Aside from our general sense of touch, the special senses include the four traditionally associated with special sensation; audition, vision, olfaction, and gustation. In addition, a fifth special sense exists, that of equilibrium. These five special senses have several characteristics in common. They are produced by concentrated and organized sense organs consisting of the specialized receptor cells and associated neurons. They are also all designed to examine conditions in the external environment and respond to either chemical, photoelectric, or mechanical energy.

Test Yourself

1. What produces your sense of smell? Describe the olfactory receptors.

2. Why does sniffing intensify the sense of smell?

3. What are primary scents? What adaptation occurs to odors? What is odor threshold?

4. Describe the pathway for olfaction as impulses from the olfactory receptors rise to become perceptions in the CNS.

5. What produces your sense of taste? Describe the gustatory receptors.

6. Label the following diagram of the taste zones of the tongue and parts of a taste bud.

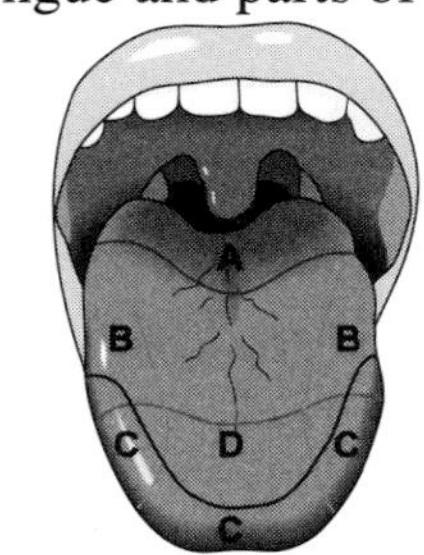

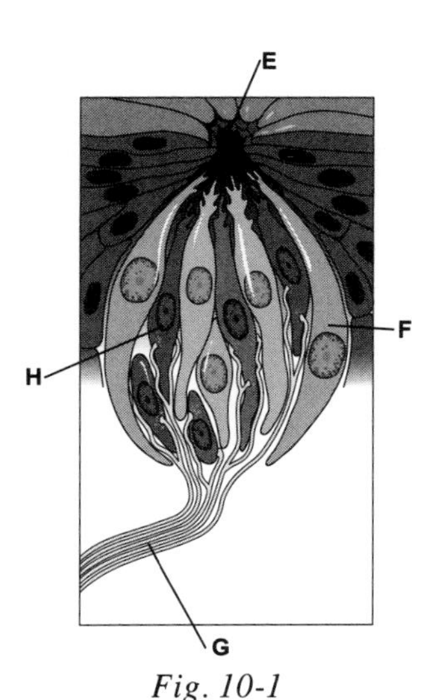

Fig. 10-1

7. What are the primary tastes? Why are we sensitive to these specific tastes?

8. What produces stimulation of gustatory receptors? What adaptation occurs to tastes? What is the taste threshold?

9. Describe the pathway for gustation as impulses from the gustatory receptors rise to become perceptions in the CNS.

Questions 10–24 are matching. Match the following accessory structure of the eye with its definition.

10. Ducts that drain the lacrimal fluid from the surface of the eye into the nasolacrimal sac — a. Palpebra

11. Thin mucous membrane of primarily stratified columnar epithelial cells and goblet cells that acts to protect the cornea of the eye — b. Tarsal Plate

12. Carries lacrimal fluid from the nasolacrimal sac to its release in the nasal cavity — c. Tarsal glands

13. Composed of epidermis, dermis, subcutaneous tissue, fibers of the orbicularis oculi muscle, tarsal plate, tarsal glands, and conjunctiva; the eyelid — d. Conjunctiva

14. Short hairs that project from the border each eyelid; serve to protect the eye from foreign objects, perspiration, and the direct rays of the sun — e. Eyelashes

15. Small openings that allow lacrimal fluid to drain from the eye and enter the lacrimal canals — f. Eyebrows

16. Watery solution containing salts, mucus, and a bactericidal lysozyme that cleans and lubricates the surface of the eye — g. Glands of Zeis

17. Empties lacrimal fluid from the lacrimal glands onto the conjunctiva of the upper palpebrae and onto the surface of the eye

18. Structure that serves to hold recovered lacrimal fluid prior to its release to the nasal cavity

19. Sebaceous ciliary glands that produce a lubricating fluid for the eyelashes

20. Hairs that project from the skin covering the supraorbital ridge of the frontal bone forming an arch that helps to protect the eye from foreign objects and perspiration

21. Thick band of connective tissue to produce form and support for palpebrae

22. A group of structures that produces, delivers, and drains lacrimal fluid

23. Also known as Meibomian glands, secrete oil that prevents palpabra from adhering to one another

24. Produces and secretes lacrimal fluid into the excretory lacrimal ducts

h. Lacrimal apparatus

i. Lacrimal fluid

j. Lacrimal glands

k. Excretory lacrimal ducts

l. Lacrimal puncta

m. Lacrimal canals

n. Nasolacrimal sac

o. Nasolacrimal duct

25. Label the following diagram of a transverse section through the eyeball.

26. Describe the cavities and chambers of the eye and identify the humors found in each. How are the humors produced and what are their functions?

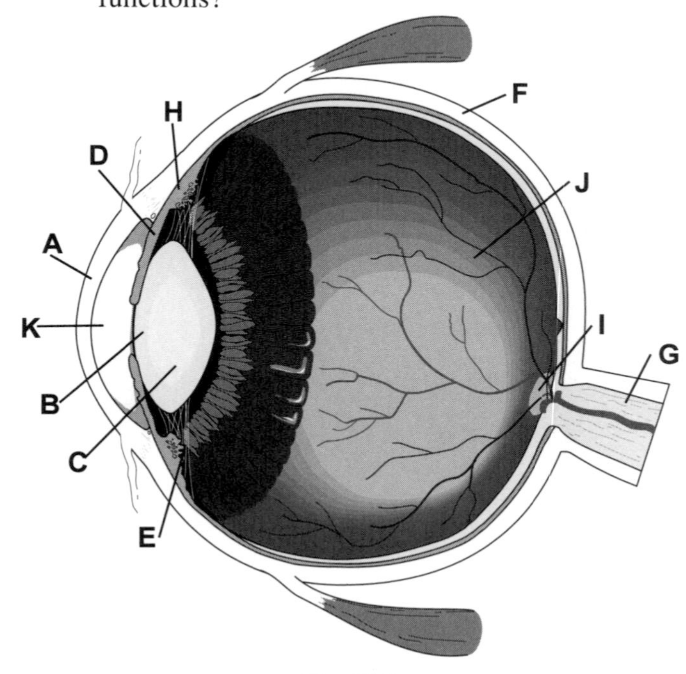

Fig. 10-2

27. Label the following diagram of the histology of the retina.

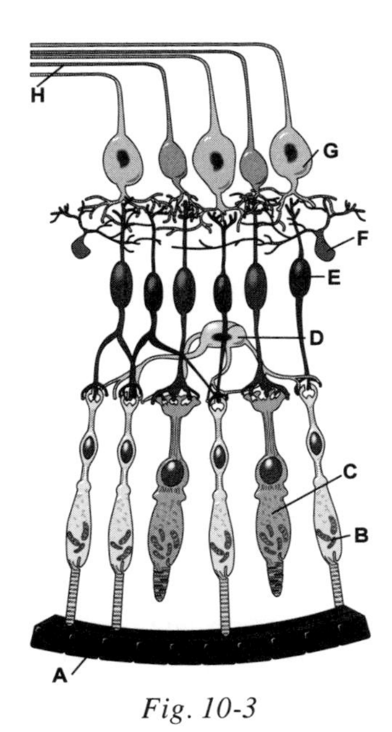

Fig. 10-3

28. Describe the structure of rods and cones.

29. Explain how photopigments respond to light. How do receptor potentials occur?

30. Explain the following as it relates to vision:

    a. refraction of light

    b. accomodation

    c. constriction of the pupil

    d. convergence

31. What is the macula lutea? What is the central fovea? What is their functional significance? What causes a "blind spot"?

32. Describe the visual fields and pathway of visual information as impulse rise from the retina to become perceptions in the CNS.

33. Label the following diagram of the outer, middle, and inner ear.

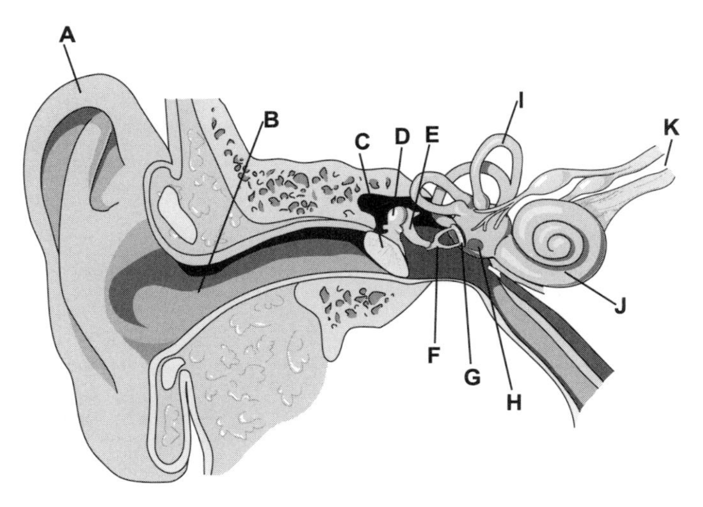

Fig. 10-4

34. What are sound waves? What are the characteristics of sound waves and how are they measured (expressed)?

35. Describe the pathway and transmission of sound waves (energy) from the auricle to the organ of Corti.

36. Label the following diagram of the organ of Corti.

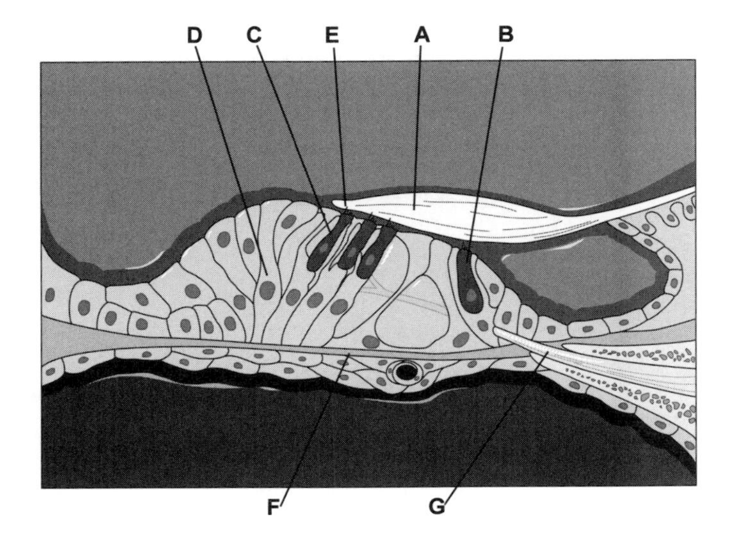

Fig. 10-5

37. Describe a hair cell. How do hair cells convert mechanical energy into an electrical signal?

38. How are the perceptions of pitch and loudness (intensity) produced by the responses of hair cells in the organ of Corti?

39. Describe the pathway of auditory information as impulses from the organ of Corti arise to perception in the CNS.

40. Describe the structure of the semicircular canals. What is the function of the semicircular canals and how is this accomplished?

41. Describe the structure of the vestibule. What is the function of the vestibule and how is this accomplished?

42. Describe the pathway of equilibrium information as impulses from the vestibule and semicircular canals arise to perception in the CNS.

Check Yourself

1. The sense of smell is produced by the interaction of chemicals with olfactory receptors. The olfactory receptor cells are first-order sensory neurons, bipolar in structure with a knob shaped dendrite that has olfactory hairs (cilia) that project into the mucus layer of the nasal cavity. The axon of the olfactory receptor cell projects to the olfactory bulb. **(Olfactory receptor cells)**

2. Olfactory receptors are housed in the olfactory epithelium located on the roof of the nasal cavity. In order for air to reach the olfactory epithelium as it enters the nasal cavity, it must make a sharp turn above the middle nasal concha. Sniffing draws the air more superiorly, increasing the flow of air over the olfactory epithelium and intensifying the reception of the odor. **(Olfactory system anatomy)**

3. Primary scents are thought to be specific classes of chemicals to which the olfactory cells will differentially respond. Despite many efforts, the odorants which represent primary scents have remained elusive (there may be hundreds). Some odorants are known to bind to receptors in the plasma membrane surrounding the olfactory cells resulting in the opening of chemically gated Na^+ channels. Olfactory receptors have a rapidly decreasing sensitivity (adaptation) to stimuli, as much as 50 percent in the first second of exposure. The threshold for olfaction is quite low, usually only requiring a few molecules for sensation. **(Olfaction)**

4. The axons of the olfactory cells produce the olfactory (I) nerves. These project to the olfactory bulb where they synapse with second-order neurons. The axons of these olfactory bulb neurons form the olfactory tract (cranial nerve) that projects to the lateral olfactory area of the temporal lobe. **(Olfactory tract)**

5. The sense of taste is also produced by the interaction of chemicals with sensory receptors. The chemicals must be dissolved for the interactions to occur. Gustatory receptor cells are specialized epithelial cells associated with fibers from a cranial nerve. These cells have a single gustatory hair (microvillus) that projects into the oral cavity. **(Gustatory receptor cells)**

6. a. Bitter zone

   b. Sour zone

   c. Salty zone

   d. Sweet zone

e. Taste pore

f. Supporting cell

g. Fibers of a cranial nerve

h. Gustatory receptor cell **(Gustatory system anatomy)**

7. The primary tastes represent four major classes of chemicals to which our gustatory receptor cells will differentially respond. Complex tastes are combinations of the four primaries found in different ratios in our food or drink. Our sensitivity to the four primary tastes; salty, sweet, sour, and bitter, represent our requirements of certain substances. Our taste for salt helps to maintain our electrolyte balance, our taste for sweet helps to provide for our requirement for carbohydrates and some amino acids, and our taste for sour helps to identify sources of vitamin C and other vitamins. Our distaste for bitter substances is a protective response in that many of the naturally occurring plant toxins and spoiled foods produce this taste. **(Gustation)**

8. Gustatory receptor cells, like olfactory receptor cells, are thought to produce binding sites for specific chemicals and when binding occurs, generate receptor potentials in the cranial nerve fibers with which they associate. Adaptation also occurs in taste reception, although not as rapidly. The threshold varies among the four primary tastes with bitter as the most sensitive, followed by sour. Salty and sweet thresholds are about the same and slightly higher than sour. **(Gustation)**

9. Three different cranial nerves supply the first-order neurons to the taste buds: the facial, glossopharyngeal, and vagus. Impulses are conducted to the nuclei of these nerves in the medulla oblongata and then on to the hypothalamus or thalamus. From the thalamus, pathways project to the primary gustatory area of the parietal lobe. **(Gustatory tract)**

10. m **(Cavities and humors of the eye)**

11. d **(Cavities and humors of the eye)**

12. o **(Cavities and humors of the eye)**

13. a **(Cavities and humors of the eye)**

14. e **(Cavities and humors of the eye)**

15. l **(Cavities and humors of the eye)**

16. i **(Cavities and humors of the eye)**

17. k **(Cavities and humors of the eye)**

18. n **(Cavities and humors of the eye)**

19. g **(Cavities and humors of the eye)**

20. f **(Cavities and humors of the eye)**

21. b **(Cavities and humors of the eye)**

22. h **(Cavities and humors of the eye)**

23. c **(Cavities and humors of the eye)**

24. j **(Cavities and humors of the eye)**

25. a. Cornea

    b. Pupil

    c. Lens

    d. Iris

    e. Suspensory ligaments

    f. Sclera

    g. Optic nerve

    h. Ciliary body

    i. Optic disk

    j. Posterior cavity

    k. Anterior cavity **(Anatomy of the eye)**

26. The eye is composed of two cavities, the anterior cavity which is filled with aqueous humor and the posterior cavity, which is filled with vitreous humor (the vitreous body). These cavities are effectively separated by the lens, ciliary body, and suspensory ligaments. The anterior cavity is further divided into the anterior and posterior chambers, with the iris serving as the division. The aqueous humor is produced as a filtrate of the blood by capillaries in the ciliary process filling the posterior chamber and flowing through the pupil to fill the anterior chamber. It is drained by the canal of Schlemm and returned to the blood. The aqueous humor nourishes the lens and cornea and produces intraocular pressure responsible for helping to maintain the shape of the eye. The posterior cavity is filled with the vitreous body, a gelatinous substance formed during embryonic life that also serves to maintain the shape of the eyeball. **(Cavities and humors of the eye)**

27. a. Pigmented epithelium

    b. Rod

    c. Cone

    d. Horizontal cell

e. Bipolar cell

f. Amacrine cell

g. Ganglion cell

h. Optic nerve **(Retina anatomy)**

28. Rods and cones are specialized photoreceptor neurons (generally bipolar in structure) that include an outer segment (a highly specialized dendrite) that contains photopigments embedded in membrane folds or discs. The inner segment contains the soma of the cell, including the nucleus, mitochondria, and Golgi apparatus in particular. The axon projects to synapse with bipolar and horizontal cells. **(Photoreceptor cell anatomy)**

29. Visual transduction begins with the absorption of light energy by the photopigments of the outer segment. These are pigmented proteins that undergo structural changes in response to light absorption. Rods contain rhodopsin, a single photopigment, while cones contain one of three different kinds of photopigments which respond to one of the primary colors. All the photopigments contain a glycoprotein called opsin and a derivative of vitamin A called retinal. Retinal is the light absorbing part of the photopigment. Small variations in amino acid sequences produce the four different opsins. In darkness, retinal assumes a particular shape, cis-retinal, that associates with opsin. With the absorption of light, cis-retinal is converted to the isomer trans-retinal and disassociates from opsin. When light is again absent, the enzyme retinal isomerase assists in the conversion back to cis-retinal and the association to opsin. In darkness, Na^+ ions flow into the outer segment through channels held open by cyclic guanosine monophosphate (c-GMP). This continual depolarization produces a continuous release of neurotransmitter by the photoreceptor cells, that inhibit the activity of the post-synaptic bipolar cells. Bipolar cells, like the rods and cones, are tonically active unless inhibited. When light is absorbed by the retinal, enzymes are activated that serve to break down c-GMP and close the Na^+ channels, preventing the release of the inhibitory influence on bipolar neurons, allowing for their activation. **(Physiology of photoreceptors)**

30. a. Refraction of light occurs as light passes through a transparent substance into a second transparent substance with a density different from the first, producing a bending of the light ray. As light enters the eye, it is refracted slightly at both surfaces of the cornea and more dramatically at both surfaces of the lens.

    b. Accomodation is the increase in the curvature of the lens. The bi-convex nature of the lens serves to bend the rays of light toward each other, allowing for a juncture of the light rays at some point (focusing on the central fovea of the retina). By increasing the curvature of the lens, light rays can be bent more to allow for focusing on near objects.

    c. The circular muscles of the iris contract to constrict the size of the opening of the pupil. This automatic reflex occurs to restrict the amount of light entering the eye in bright light, but also acts to prevent light from entering the periphery of the lens. Acting in concert with accommodation, this prevents light that would not be in focus from reaching the retina resulting in clearer vision.

    d. Convergence refers to the medial movement of both eyes in order to maintain single binocular vision on close objects. **(Image formation)**

31. The macula lutea is the center of the posterior portion of the retina. The central fovea is a slight depression in the retina at the center of the macula lutea produced by a slightly thinner retina at this point. Here the retina contains only cones, with the associated bipolar and ganglion cells displaced to the side.

This allows for the highest visual acuity in the center of the visual field in bright light. The blind spot is the point at which the optic nerve exits the eye, a place where no photoreceptor cells exist, thus a place where light cannot be processed. **(Anatomy of the retina)**

32. The output of the retina is through the axons of the ganglion cells which form the cranial optic nerve (optic nerve II fibers). Most of the fibers cross at the optic chiasm as they enter the optic tract. The optic tract projects to the synapse on neurons of the lateral geniculate of the thalamus. These neurons project to the primary visual cortex in the occipital lobe. **(Optic tract)**

33. a. Auricle

    b. External auditory canal

    c. Tympanic membrane

    d. Malleus

    e. Incus

    f. Stapes

    g. Oval window

    h. Semi-circular canals

    i. Vestibule

    j. Cochlea

    k. Vestibulocochlear nerve **(Anatomy of the auditory system)**

34. Sound waves are the alternate compression and decompression of air molecules. They originate with vibrating objects. The two characteristics of sound are frequency (pitch) and intensity (loudness). Frequency of sound is a result of the wavelengths of sound measured in the rapidity of their occurrence in space. Sound frequency is typically measured in cycles per second. Intensity of sound is an indication of the amount of energy contained in each wave measured in units called decibels with normal speech having a level of approximately 60 dB. **(Characteristics of sound)**

35. Sound waves are directed by the auricle into the external auditory canal where they produce vibrations of the tympanic membrane. The tympanic membrane is attached to the ossicular chain. The vibrations of the tympanic membrane are transferred and amplified by the malleus, incus, and stapes to the oval window. The oval window is a membrane similar in structure to the tympanic membrane. As the stapes vibrates against the oval window, waves are formed in the perilymph fluid filling the cavity of the scala vestibuli that connects to the oval window.

    The scala vestibuli is the upper chamber of the cochlea and is continuous with the lower chamber, the scala tympani, by the helicotrema at the apex of the cochlea. The waves travel through the perilymph of the scala vestibuli through the helicotrema into the scala tympani. The waves traveling through the scala tympani produce vibrations in the basilar membrane of the organ of Corti that forms the roof of the scala tympani. It is here that the auditory receptor cells are activated. **(Anatomy of the auditory system)**

36. a. Tectorial membrane

    b. Inner hair cells

    c. Outer hair cells

    d. Supporting cells

    e. Microvilli (hairs)

    f. Basilar membrane

    g. Vestibulocochlear nerve **(Anatomy of the organ of corti)**

37. Hair cells are located atop the basilar membrane and have long, hair-like microvilli at their apical ends that extend into the endolymph of the cochlear duct with their ends against the tectorial membrane. The basal ends of these cells synapse with first-order auditory neurons. The mechanical deformation (bending) of the microvilli as they are pressed against the tectorial membrane is responsible for opening ion channels in the plasma membrane and producing a depolarization. **(Anatomy of the hair cells)**

38. Different intensities of sound waves will produce different amounts of energy in the waves of the perilymph of scala of the cochlea. The greater the energy, the greater the displacement of the basilar membrane and thus the greater amount of bending of the microvilli of the hair cells, resulting in a faster firing frequency of the responding neurons. Different frequencies of sound produce vibrations most efficiently at different locations along the basilar membrane such that different populations of hair cells located in the specific areas of vibrations are activated. **(Physiology of auditory receptor cells)**

39. The first order sensory neurons project along the vestibulocochlear (VIII) nerve to carry impulses to the cochlear nuclei in the medulla oblongata. Neurons from the cochlear nuclei project to the thalamus. Thalamic neurons project to the primary auditory cortex in the temporal lobe. **(Auditory tract)**

40. The semicircular canals are three bony labyrinths lying at right angles to one another in three planes. The two vertical ones are the anterior and posterior semicircular canals, respectively and the horizontal one is the lateral semicircular canal. Each is filled with perilymph and contains a membranous labyrinth filled with endolymph. At the base of each is an ampulla containing a small elevation called the crista which has a group of hair cells and supporting cells. The microvilli of the hair cells are embedded in a gelatinous matrix called the cupula. Movements of the head (accelerations) cause the endolymph to flow against the cupula bending the microvilli of the hair cells. This provides for a sense of dynamic equilibrium in any of the three primary directions the body can move. **(Semicircular canals and dynamic equilibrium)**

41. The vestibule is the bony, central labyrinth of the inner ear. Its membranous labyrinth contains two sacs, the utricle and saccule, which are at roughly right angles to one another. The walls of these sacs contain a thickened region called the macula. Each macula consists of a layer of hair cells and supporting cells. The hair cells have long membrane extensions called stereocilia (microvilli) and one conventional cilium called a kinocilium.

    A thick, gelatinous matrix of glycoproteins with embedded calcium carbonate crystals (otoliths) rests upon the hair cells known as the otolithic membrane. When the head is tilted with respect to gravitational field, this heavy, otolithic membrane will slide, bending and activating the hair cells. The action of the saccule and utricle is primarily responsible for static equilibrium. **(Vestibule and static equilibrium)**

42. The hair cells of the vestibule and semicircular canals activate first order neurons of the vestibulocochlear (VIII) nerve which project to the vestibular nuclei in the pons or to the cerebellum through the inferior cerebellar peduncle. Neurons from these vestibular nuclei project to the nuclei that control eye movements: oculomotor, trochlear, and abducens. They also project to the nuclei of the accessory nerve to assist in the control of head and neck movements. **(Equilibrium tract)**

Grade Yourself

Circle the numbers of the questions you missed, then fill in the total incorrect for each topic. If you answered more than three questions incorrectly, you need to focus on that topic. (If a topic has less than three questions and you had at least one wrong, we suggest you study that topic also. Read your textbook, a review book, or ask your teacher for help.)

Subject: The Special Senses

Topic	Question Numbers	Number Incorrect
Olfactory receptor cells	1	
Olfactory system anatomy	2	
Olfaction	3	
Olfactory tract	4	
Gustatory receptor cells	5	
Gustatory system anatomy	6	
Gustation	7, 8	
Gustatory tract	9	
Cavities and humors of the eye	10, 11, 12, 13, 14, 15, 16, 17, 18, 19, 20, 21, 22, 23, 24, 25, 26	
Retina anatomy	27	
Photoreceptor cell anatomy	28	
Physiology of photoreceptors	29	
Image formation	30	
Anatomy of the retina	31	
Optic tract	32	
Anatomy of the auditory system	33, 35	
Characteristics of sound	34	
Anatomy of the organ of corti	36	
Anatomy of the hair cells	37	
Physiology of auditory receptor cells	38	
Auditory tract	39	
Semicircular canals and dynamic equilibrium	40	
Vestibule and static equilibrium	41	
Equilibrium tract	42	

The Endocrine System

Brief Yourself

The endocrine system works in concert with the nervous system in organizing the body, providing regulation and synchronicity among the various structures and processes. In contrast to the nervous system, the endocrine system acts more slowly and typically produces longer lasting results. The actions of this system provide for regulation of growth, development of body tissues and tissue functions, and a determination of the rate of many metabolic processes of the body.

The endocrine system is composed of the endocrine glands of the body. The endocrine cells of these glands produce and secrete hormones into the extracellular fluid surrounding the cells. These hormones enter the circulatory system and are distributed throughout the body where they will bind to appropriate receptors, affecting metabolic regulation. In addition to the specific glands of the endocrine system, certain organs also contain cells with endocrine function that generally regulate the functions of these organs.

Test Yourself

1. Label the glands in the diagram below.

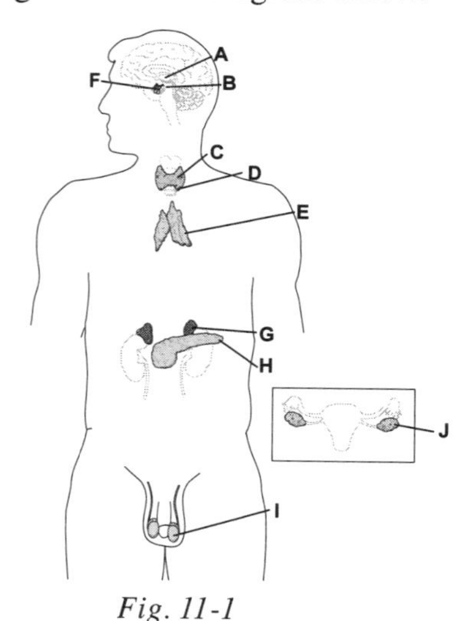

Fig. 11-1

2. What are hormones?

3. What are the two basic mechanisms of hormone function and how do they differ?

Questions 4–8 are matching. Match the following hormone with the appropriate gland.

4.	Thyroid	a.	Parathormone
5.	Pineal	b.	Epinephrine
6.	Posterior pituitary	c.	Calcitonin
7.	Parathyroid	d.	Melatonin
8.	Adrenal medulla	e.	Oxytocin

9. What part of the brain directly influences endocrine function? What is the mechanism of neuroendocrine communication?

Questions 10–17 are matching. Match the appropriate pituitary hormone with its principal effect.

10.	GH	a.	Stimulates contraction of smooth muscle of uterus during parturition
11.	FSH	b.	Controls secretion of glucocorticoids
12.	TSH	c.	Promotes milk secretion
13.	LH	d.	Conserves body water by decreasing urine volume
14.	PRL	e.	Stimulates secretion of estrogen and progesterone in females; stimulates development of testosterone in males
15.	ACTH	f.	Stimulates secretion of thyroid hormones
16.	OT	g.	Stimulates growth of body tissues
17.	ADH	h.	Stimulates development of oocytes in females; stimulates production of sperm in males

18. The thyroid gland produces two basic types of hormones, thyroxine and calcitonin. What is the effect of each?

19. Identify the primary secretion of the parathyroid gland. What is its principal effect?

20. Identify the three primary secretions of the adrenal cortex. What is the principal effect of each?

21. Identify the two primary secretions of the adrenal medulla. How does the effect of these hormones relate to the autonomic nervous system?

22. The endocrine role of the pancreas is the production of glucagon and insulin. How do these hormones interact to regulate blood glucose levels?

23. How does the amount of sunlight experienced by an individual affect the pineal gland and its production of melatonin?

24. What are eicosanoids? How are they released?

Questions 25–33 are matching. Match the following hormone produced by organs that contain endocrine cells with their principal effects.

25.	Gastrin	a.	Stimulates production and release of pancreatic juice and bile
26.	Secretin	b.	Decreases blood pressure
27.	GIP	c.	Aids in the absorption of calcium and phosphorus
28.	CCK	d.	Increase red blood cell production
29.	hCG	e.	Inhibits secretion of gastric juice
30.	hCS	f.	Promotes secretion of gastric juice
31.	EPO	g.	Stimulates secretion of pancreatic juice and brings about a feeling of satiety (fullness) after a meal

32. Calcitrol h. Stimulates corpus luteal production of estrogen and progesterone to maintain pregnancy

33. ANP i. Stimulates the development of mammary glands

34. Identify the three concentric layers of the adrenal cortex and the class of secretion it produces.

35. Label the following diagram of a thyroid follicle.

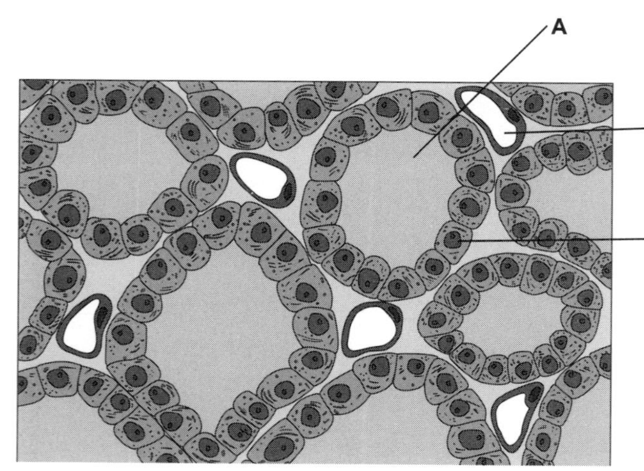

Fig. 11-2

36. Label the following diagram of an Islet of Langerhans.

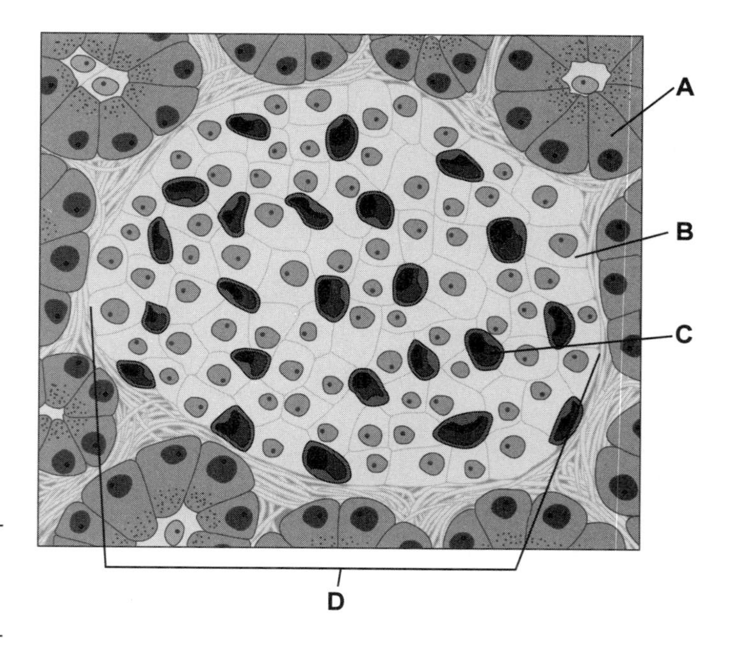

Fig. 11-3

Check Yourself

1. a) pineal gland

   b) hypothalamus

   c) thyroid gland

   d) parathyroid glands

   e) thymus gland

   f) pituitary gland

   g) adrenal glands

   h) pancreas

   i) testes

   j) ovaries **(Physiology of the endocrine system)**

2. Hormones are chemicals secreted by endocrine cells which will alter the physiology of target cells. Hormones that pass into the blood and bind to receptors on distant target cells are called circulating hormones. Those that act locally without entering the bloodstream are called local hormones. **(Hormones)**

3. Hormones are either lipid-soluble or water-soluble. Lipid-soluble hormones diffuse directly through the phospholipid bilayer of the plasma membrane of cells to bind with and activate receptors located in the cytosol. The activated receptors will in turn alter gene expression in that cell, altering its physiology. Water-soluble hormones bind with receptors embedded in the plasma membrane of cells, that activates a globular protein of the membrane that activates adenylate cyclase. Adenylate cyclase converts ATP into cyclic AMP in the cytosol. Cyclic AMP acts as the second messenger to activate protein kinase which in turn will phosphorylate additional enzymes altering the physiology of the target cell. **(Hormones)**

4. c **(Hormones)**

5. d **(Hormones)**

6. e **(Hormones)**

7. a **(Hormones)**

8. b **(Hormones)**

9. The hypothalamus of the brain is the most directly responsible for neural influence on the endocrine system. The functional connection between the hypothalamus and the pituitary gland provides the largest single component of the neuroendocrine system. The hypothalamus contains neurosecretory cells which produce and deliver hormones to the pituitary gland. The neurohypophysis, or posterior lobe of the pituitary gland, stores and releases two hormones that are synthesized by neurosecretory cells located in the hypothalamus.

   The axons of these cells project to the neurohypophysis through the infundibulum where the release of the hormones occurs. The endocrine cells of the adenohypophysis, or anterior lobe of the pituitary gland, synthesize and release hormones, but under the control of the hypothalamus. The hypothalamus produces a group of releasing or inhibiting hormones that are delivered to the adenohypophysis to regulate the release of the hormones produced by the endocrine cells of the adenohypophysis. **(Physiology of the endocrine system)**

10. g **(Hormones)**

11. h **(Hormones)**

12. f **(Hormones)**

13. e **(Hormones)**

14. c **(Hormones)**

15. b **(Hormones)**

16. a **(Hormones)**

17. d **(Hormones)**

18. Thyroxine is active in elevating the basal metabolic rate. It results in the increased synthesis of proteins, increased utilization of glucose for cellular respiration, and increased lipolysis. Calcitonin lowers the blood levels of calcium and phosphates by inhibiting their uptake in the digestive system and increasing the activity of osteoblasts. **(Physiology of the endocrine system)**

19. Parathormone works in opposition to calcitonin by increasing the calcium and magnesium content of the blood. It enhances the activity of osteoclasts and increases the dietary uptake of calcium and magnesium. **(Physiology of the endocrine system)**

20. The adrenal cortex produces mineralocorticoids, glucocorticoids, and androgens. Mineralocorticoids, particularly aldosterone, help to control water and electrolyte homeostasis, particularly with respect to sodium and potassium ions. Glucocorticoids, particularly cortisol, regulate metabolism in response to stress. Glucocorticoids increase protein catabolism, increase glucose formation, increase lipolysis, inhibit inflammation, and generally depress immune respones. Androgens are male sex hormones that have masculizing effects. **(Physiology of the endocrine system)**

21. The adrenal medulla produces epinephrine and norepinephrine. These are both sympathomimetic hormones which act to enhance the effects of the sympathetic division of the ANS. **(Physiology of the endocrine system)**

22. Low blood glucose levels stimulates the release of glucagon from the alpha cells of the pancreas. Glucagon accelerates the activity of hepatocytes of the liver, producing glycogenolysis and gluconeogenesis. The release of glucose to the bloodstream by the hepatocytes causes the levels to rise. If blood levels continue to rise, beta cells of the pancreas are activated and release insulin, producing an acceleration of the facilitated diffusion of glucose into the cells, particularly in muscle tissue, as well as accelerating glycogenesis, resulting in a decrease in glucose levels in the blood. **(Physiology of the endocrine system)**

23. When light enters the eye and stimulates the photoreceptors, impulses from the photoreceptors are transmitted to the suprachiasmatic nucleus of the hypothalamus. From here, the hypothalamic neurons project to the superior cervical ganglion which form connections to the pineal gland. When these neuron circuits are more active, during periods of increased sunlight, the production of melatonin is inhibited assisting in arousal. During periods of decreased sunlight, melatonin production increases producing drowsiness. **(Physiology of the endocrine system)**

24. Eicosanoids are local hormones that are produced in all tissues of the body. They produce a broad range of physiological effects and are often released by cells in response to local tissue damage. The two basic classes of eicosanoids are prosglandins and leukotrienes. **(Physiology of the endocrine system)**

25. f **(Hormones)**

26. a **(Hormones)**

27. e **(Hormones)**

28. g **(Hormones)**

29. h **(Hormones)**

30. i **(Hormones)**

31. d **(Hormones)**

32. c **(Hormones)**

33. b **(Hormones)**

34. The cortical cells of the adrenal cortex are arranged in three layers. The most superior layer is the zona glomerulosa which produces mineralocorticoid hormones. The middle layer is the zona fasciculata which produces glucocorticoids and the most profound layer is the zona reticulata which produces the goadocorticoids. **(Physiology of the endocrine system)**

35. a) thyroglobulin

    b) blood vessel

    c) follicular cell **(Physiology of the endocrine system)**

36. a) Acinar cell

    b) Alpha cell

    c) Beta cell

    d) Islet of Langerhans **(Physiology of the endocrine system)**

Grade Yourself

Circle the numbers of the questions you missed, then fill in the total incorrect for each topic. If you answered more than three questions incorrectly, you need to focus on that topic. (If a topic has less than three questions and you had at least one wrong, we suggest you study that topic also. Read your textbook, a review book, or ask your teacher for help.)

Subject: The Endocrine System

Topic	Question Numbers	Number Incorrect
Physiology of the endocrine system	1, 9, 18, 19, 20, 21, 22, 23, 24, 34, 35, 36	
Hormones	2, 3, 4, 5, 6, 7, 8, 10, 11, 12, 13, 14, 15, 16, 17, 25, 26, 27, 28, 29, 30, 31, 32, 33	

The Cardiovascular System: The Blood

Brief Yourself

The cardiovascular system is composed of the heart, the blood vessels, and the blood. It functions as a remarkably efficient transport system. The tissues of the body continuously require nutrients and oxygen and produce wastes that require excretion. Because most cells are only capable of such exchanges with the extracellular fluid of their immediate environment, a method of renewing that environment is critical in multicellular organisms. The cardiovascular system provides for a steady supply of nutrients to the tissues and a removal of wastes from the tissues to prevent the toxic pollution of those environments.

The transport medium for the cardiovascular system is the blood. Blood is a connective tissue and is the only fluid tissue of the body. Blood is thicker and more viscous than water and appears relatively homogenous, but actually is composed by formed elements, including living blood cells, suspended in a non-living fluid matrix called plasma. Blood is slightly alkaline and its temperature is typically slightly greater than body temperature. Due to its high concentration of dissolved ions, it has a slightly saline taste. Blood is opaque and red in color due to the hemoglobin content in erythrocytes (red blood cells). It varies from a bright scarlet color in oxygenated blood to a dull, dark red in deoxygenated blood.

Test Yourself

1. How is the total blood volume of a person regulated and maintained?

2. Identify the blood cells in Figure 12-1.

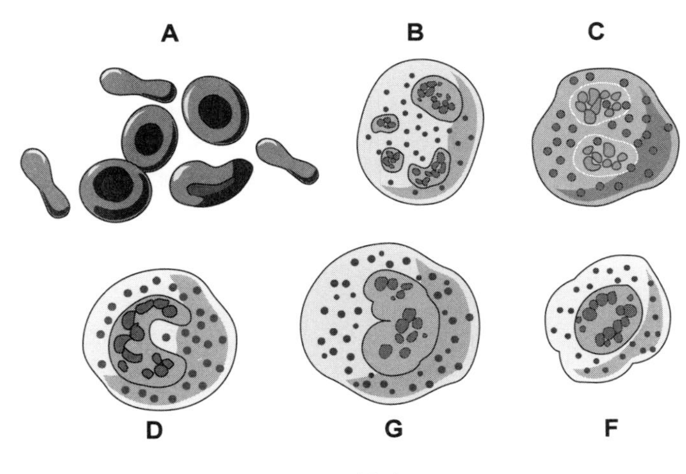

Fig. 12-1

3. Discuss how a platelet plug forms.

4. What are the chemical structure and function of hemoglobin? How is it transported?

5. What are the components of the blood? If a blood sample is treated with an anti-coagulant and spun in a centrifuge what three layers are typically seen and what percentage of whole blood is represented by each?

6. What factors influence the viscosity of the blood?

7. What is hematopoiesis and where does it occur?

Questions 8–14 are matching. Please match the following formed element of the blood with its function.

8. Erythrocyte	a. Transforms to macrophage; phagocytosis
9. Neutrophil	b. Liberates heparin, histamine, and serotonin in allergic reactions to intensify the inflammatory effect
10. Eosinophil	c. Contains hemoglobin and transports oxygen and carbon dioxide
11. Basophil	d. Function in hemostasis; release chemicals to enhance vascular spasm
12. Lymphocyte	e. Phagocytosis of bacteria; destruction of bacteria through utilization of defensins, lysozyme, or oxidants
13. Monocyte	f. Opposes effect of histamine in allergic response and phagocytize antigen-antibody complexes; destroys certain parasitic worms
14. Platelets	g. Mediate immune responses; natural killer cells, B-cells, and T-cells

15. What is the difference between the extrinsic and intrinsic pathways to blood clotting?

16. What characteristic of erythrocytes produces the ABO blood group? How does this effect whole blood transfusions?

Questions 17–20 are matching. Please match the following ABO blood type with compatible donor types.

17. A
18. B
19. AB
20. O

a. O
b. A, B, AB, O
c. B, O
d. A, O

21. In the Rh blood group, why is it likely that an Rh- mother can deliver a firstborn that is Rh+ with no ill effects, but a second fetus that is Rh+ will typically suffer hemolysis of the fetal blood due to the mother's immune response?

22. Describe the anatomy of an erythrocyte. Why is the life span of these cells limited to approximately 120 days?

23. Leukocytes exhibit diapedesis and chemotaxis. What are these characteristics?

24. How are platelets produced?

Check Yourself

1. Blood volume is regulated and maintained by actions of the several negative feedback systems and the capability of coagulation during trauma to the vascular system. Particularly important is the regulation of water volume involving anti-diuretic hormone, aldosterone, and atrial natriuretic peptide. These hormones influence the function of the kidney and thus influence the loss or retention of water. **(Physiology of the blood)**

2. a) erythrocyte

   b) neutrophil

   c) basophil

   d) eosinophil

   e) monocyte

   f) lymphocyte

   g) monocyte **(Anatomy of blood cells)**

3. Platelet plug formation is a component of hemostasis. Platelet plugs typically form on the inner walls of blood vessels in response to trauma. Circulating blood deposits platelets at the injury site that adhere to the collagen fibers uncovered by the damage to the endothelial lining of the vessel. As a result of adhesion, platelets begin to swell and release the contents of their granules, in particular, ADP. ADP increases the "stickiness" of other platelets in the area, causing additional adhesion to the originally activated platelets. This aggregation of platelets reduces blood loss at the point of the injury by forming a mass called the platelet plug. **(Hemostasis)**

4. Hemoglobin is carried in the erythrocytes. Each erythrocyte will contain approximately 280 million hemoglobin molecules. A hemoglobin molecule consists of four polypeptide chains, each bound to an iron containing heme group and each heme group will reversibly bind and carry one molecule of oxygen. Hemoglobin carries about 98.5 percent of all the oxygen transported in the cardiovascular system as well as transporting about 23 percent of the carbon dioxide that binds to amino acids of the polypeptide chains forming carbaminohemoglobin. **(Physiology of the blood)**

5. Blood is composed of plasma and formed elements. Centrifuged blood will form an upper layer of plasma (55 percent), a middle layer called the buffy coat composed of leukocytes and platelets (1 percent), and a lower layer of erythrocytes (45 percent) called the hematocrit. **(Anatomy of the blood)**

6. Viscosity of the blood is produced primarily as a function of the amount of formed elements in the blood. The greater the ratio of formed elements to plasma the more viscous the blood becomes. **(Anatomy of the blood)**

7. Hematopoiesis is the production of new blood cells that occurs in the hematopoietic stem cells found distributed in the red bone marrow. Erythropoiesis is the production of new erythrocytes and leukopoiesis is the production of new leukocytes and platelets. **(Hematopoiesis)**

8. c **(Physiology of the blood)**

9. e **(Physiology of the blood)**

10. f **(Physiology of the blood)**

11. b **(Physiology of the blood)**

12. g **(Physiology of the blood)**

13. a **(Physiology of the blood)**

14. d **(Physiology of the blood)**

15. The extrinsic and intrinsic pathways of blood clotting differ in several respects. The extrinsic pathway has fewer steps and occurs rapidly. It is triggered by thromboplastin (coagulation factor III) which is released from the surfaces of injured cells and leaks *into* the blood. Thromboplastin activates clotting factor VII that combines with factor X, thus activating it. Activated factor X combines with factor V to form active prothrombinase completing the pathway. The intrinsic pathway is triggered by elements contained *within* the blood. At damaged or roughened endothelial cells lining the blood vessels, collagen will often be exposed. Adhesion to the collagen by platelets will result in phospholipid release by the platelets that activates clotting factor XII. Activated XII will in turn activate factor XI, which activates factor IX. Activated IX will combine with factor VIII and platelet phospholipids to activate factor X. Activated X combines with factor V to form active prothrombinase. **(Hemostasis)**

16. Erythrocytes contain antigens (agglutinogens) as a component of their plasma membranes. The ABO blood group is based on two antigens, A and B, which are genetically determined. People whose erythrocytes have only type A antigens are said to have type A blood. Similarly, people with erythrocytes displaying only type B antigens have type B blood, people with erythrocytes displaying both A and B antigens have type AB blood, and people with erythrocytes displaying no antigens have type O. In addition to antigens on the erythrocytes, blood plasma contains antibodies (agglutinins) which react to bind to antigens. An antibody bound to an antigen forms the antigen-antibody complex and will trigger the complement factors of the blood, resulting in hemolysis of the cell displaying this complex. Type A blood plasma will contain only anti-B antibodies that bind to type B antigens, type B blood will contain only anti-A antibodies that bind to type A antigens, type AB blood will contain neither antibody, and type O blood will contain both anti-A and anti-B antibodies. In whole blood transfusion, care must be taken that the donor blood and recipient blood will not form the antigen-antibody complex. **(Blood groups)**

17. d **(Blood groups)**

18. c **(Blood groups)**

19. b **(Blood groups)**

20. a **(Blood groups)**

21. During childbirth, a small quantity of fetal blood leaks across the placenta to enter the mother's blood stream. If an Rh- mother has given birth to an Rh+ child (inherited from the father), the mother's immune system will respond by producing anti-Rh antibodies that will remain in the blood stream. With a second

pregnancy and a fetus carrying the Rh+ gene, and thus Rh antigen, the mother's antibodies can cross the placenta and enter the fetus, producing hemolysis of the fetal blood erythrocytes. **(Blood groups)**

22. Erythrocytes are relatively small cells (7-8 percent m in diameter) with a biconcave shape. Mature erythrocytes do not contain a nucleus. They also do not contain mitochondria, so they produce their energy only through anaerobic processes. They are essentially plasma membranes containing cytosol packed with hemoglobin molecules. Their brief life span is due to their lack of a nucleus. Without a nucleus, the normal "wear and tear" on the plasma membrane and other structures can not be repaired, leading to fragile membranes subject to hemolysis. **(Anatomy of blood cells)**

23. Leukocytes are mobile cells that move by amoeboid motion. The cardiovascular system serves as a conduit for their circulation, but they have the ability to escape the vessels to enter the tissues, squeezing through capillary wall fenestrations in a process known as diapedesis. Phagocytic leukocytes are attracted to chemicals released by toxins released by microbes, kinins released by damaged tissues, and some of the colony-stimulating factors. These leukocytes travel by amoeboid movement up the chemical gradient toward the source of the chemical release. **(Physiology of the blood)**

24. Platelets are produced by megakaryoblasts that are formed during leukopoiesis. These mature to megakaryocytes, huge cells that burst into fragments of granules encased in plasma membranes. **(Physiology of the blood)**

Grade Yourself

Circle the numbers of the questions you missed, then fill in the total incorrect for each topic. If you answered more than three questions incorrectly, you need to focus on that topic. (If a topic has less than three questions and you had at least one wrong, we suggest you study that topic also. Read your textbook, a review book, or ask your teacher for help.)

Subject: The Cardiovascular System: The Blood

Topic	Question Numbers	Number Incorrect
Physiology of the blood	1, 4	
Anatomy of blood cells	2, 22	
Hemostasis	3, 15	
Physiology of the blood	8, 9, 10, 11, 12, 13, 14, 23, 24	
Anatomy of the blood	5, 6	
Hematopoiesis	7	
Blood groups	16, 17, 18, 19, 20, 21	

The Cardiovascular System: The Heart

13

Brief Yourself

The heart is a four-chambered muscular pump, magnificently adapted to its task of propelling blood through the blood vessels. Not much larger than the size of an adult's fist and with an average weight of about 300g, this organ is responsible for pumping about 30 times its on weight each minute, even when you are sleeping. During activity, the heart's output increases to match the demands of the tissue for oxygen and nutrients. With an average heart rate of about 75 bpm, a normal human heart would pump well over 10,000,000 liters of blood per year.

The heart is located in the mediastinum, extending about 12 cm from the second rib to the fifth intercostal space. It lies anterior to the vertebral column and posterior to the sternum. The heart tips slightly to the left and assumes an oblique position in the thorax and will be partially obscured by the overlapping lungs.

The bulk of the tissue of the heart is the cardiac muscle, or myocardium. It is the activity of this tissue that is responsible for the contractility of the heart. Although similar to striate muscle, it has unique characteristics that provide for its effectiveness and durability.

Test Yourself

1. Label the structures of the heart in the anterior view below.

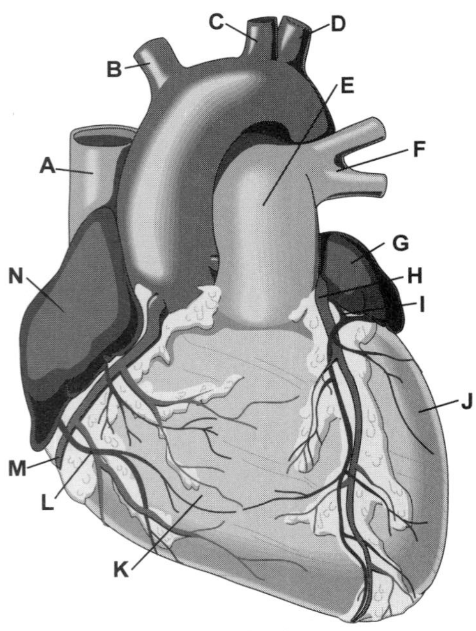

Fig. 13-1

2. Label the structures of the heart in the posterior view below.

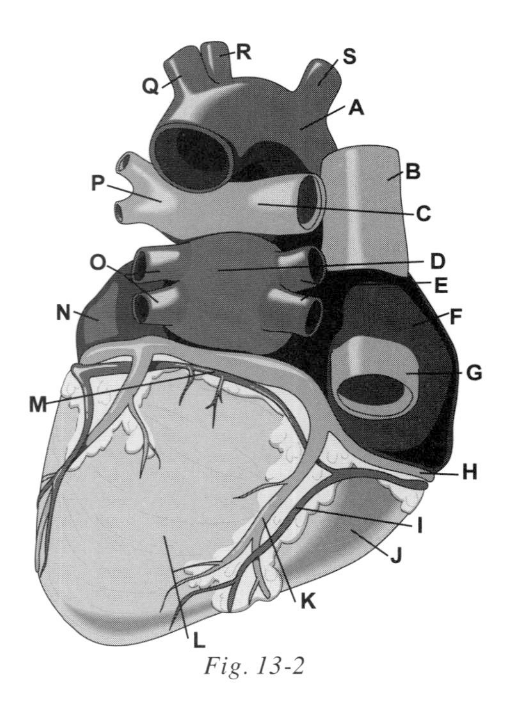

Fig. 13-2

3. Identify the four chambers of the heart. How do the functions of ventricles and atria differ?

4. Which chambers of the heart power the pulmonary circuit of blood flow? Which power the systemic circuit? How does the anatomy of these two pumping systems vary with regard to their function?

5. Identify the factors that affect heart rate.

6. Identify the two types of cardiac valves. What is their function?

7. How is cardiac output determined? What two factors most directly affect cardiac output?

8. What is the fibrous skeleton of the heart?

9. What are the layers of the pericardium that surrounds the heart? What is its function?

10. Identify the difference between skeletal muscle fiber and cardiac muscle fibers.

11. Label the diagram of the heart in the frontal section below.

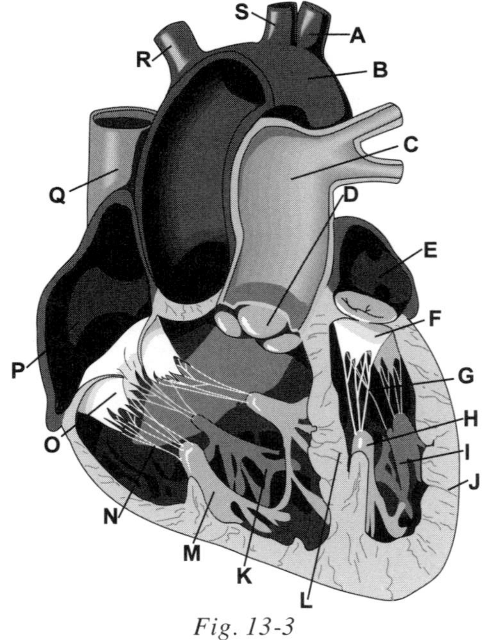

Fig. 13-3

12. Trace the pathway of blood through the heart beginning with its delivery from the vena cava.

13. What is the coronary circulation system? Why is it required and what are its major structures?

14. What are autorhythmic cells? Identify the components of the conduction system of the heart.

15. What are the basic phases that occur in the contraction of a cardiac muscle cell following the excitation by the conduction system?

16. Label the diagram of the electrocardiogram below.

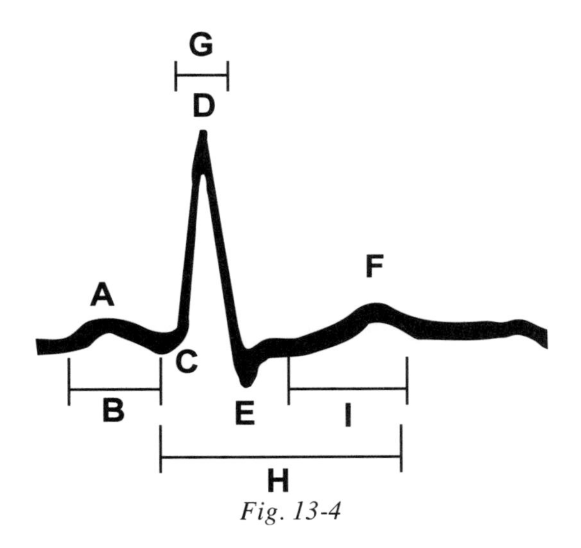

Fig. 13-4

Questions 17–22 are matching. Please match the following characteristics of an electrocardiogram with the physiological events that produce them.

17. P wave — a. represents the time when the ventricular contractile fibers are fully depolarized

18. QRS complex — b. Represents the conduction time from the beginning of atrial excitation to the beginning of ventricular excitation

19. T wave — c. Represents atrial depolarization

20. P-Q interval — d. Represents the onset of ventricular depolarization

21. S-T interval — e. Represents ventricular repolarization just before the ventricles begin to relax

22. Q-T interval — f. Represents the time from the beginning of ventricular depolarization to the end of ventricular repolarization

23. Define the terms systole and diastole. What are the three phases of the cardiac cycle?

24. What three factors affect the stroke volume of the heart?

25. What is the Frank-Starling law of the heart?

26. What is the cardiovascular center of the brain? What influences its behavior? How does it affect heart rate?

27. In addition to the activity of the autonomic nervous system, what two classes of chemical found in the blood stream have a strong effect on heart rate? What is the effect of each?

Check Yourself

1. a) Vena cava

   b) Brachiocephalic trunk

   c) Left common carotid artery

   d) Left sublavian artery

   e) Pulmonary trunk

   f) Left pulmonary artery

   g) Left auricle

   h) Left coronary artery

   i) Great cardiac vien

   j) Left ventricle

   k) Right ventricle

   l) Anterior cardiac vein

   m) Right coronary artery

   n) Right atrium **(Anatomy of the heart)**

2. a) Arch of aorta

   b) Superior vena cava

   c) Right pulmonary artery

   d) Left atrium

   e) Right pulmonary veins

   f) Right atrium

   g) Inferior vena cava

   h) Right coronary artery

   i) Middle cardiac vein

j) Right ventricle

k) Posterior interventricular branch

l) Left ventricle

m) Coronary sinus

n) Left auricle

o) Left pulmonary veins

p) Left pulmonary artery

q) Left subclavian artery

r) Left common carotid artery

s) Brachiocephalic trunk **(Anatomy of the heart)**

3. The heart is divided into four hollow chambers; the right atrium, the right ventricle, the left atrium, and the left ventricle. The atria are the superior chambers and serve as "primer" pumps to ensure the proper filling of the ventricles below them. The ventricles are the inferior, larger chambers that propel blood from the heart. **(Chambers of the heart)**

4. The right side of the heart (right atrium and right ventricle) produces blood flow to the pulmonary circuit. The left side of the heart (left atrium and left ventricle) produces blood flow to the systemic circuit. The pulmonary circuit operates at lower pressure levels than does the systemic circuit. For this reason, the walls of the right side of the heart are thinner, containing less myocardium than the left side and produce less power in its contractions. **(Chambers of the heart)**

5. Heart rate is affected by emotion, exercise, hormones, ion content of the blood, temperature, pain, and other physical stresses. **(Heart rate)**

6. The two types of cardiac valves are the atrioventricular valves and the semi-lunar valves. The atrioventricular are stronger valves with cusps of dense connective tissue anchored to the fibrous skeleton of the heart. The borders of each cusp are attached by chordae tendineae to papillary muscles. The contraction of these muscles anchor the cusps. Semi-lunar valves have similar, but smaller crescent-shaped cusps anchored to the fibrous skeleton with free borders. The function of all of the valves of the heart is the same, to direct the flow of blood by producing uni-directional flow through the valves (preventing backflow). **(Valves of the heart)**

7. Cardiac output (CO) is measured in ml/min. The formula is:

$$CO = SV \times HR$$

where SV is stroke volume (ml/beat) and HR is heart rate (beats/min). Thus, the factors affecting cardiac output are the stroke volume of the heart (ventricular volume) and the rate at which the heart beats. **(Cardiac output)**

8. The fibrous skeleton of the heart is a network of dense, fibrous connective tissue bands that gives an internal structure to the heart. It consists of two primary loops, the sinospiral and the bulbospiral. It serves to reinforce and an anchor the myocardium and valves of the heart. **(Anatomy of the heart)**

9. The pericardium is a triple-layered sac that surrounds and protects the heart in the mediastinum, yet allows sufficient freedom for the contractions and relaxations of the heart. Its outermost layer is fibrous pericardium, composed of tough, inelastic, dense irregular connective tissue. The inner serous pericardium is composed of an outer parietal layer fused to the fibrous pericardium, but composed of more elastic and finer connective tissues. The inner visceral layer is called the epicardium that adheres tightly to the surface of the heart. Between the serous layer and the epicardium is a space filled with pericardial fluid that reduces the friction between the membranes as the heart moves. **(Pericardium)**

10. Cardiac muscle fibers resemble skeletal muscle fibers primarily due to the presence of striations, although not as rigidly organized as those of skeletal muscle. Cardiac muscle fibers differ from skeletal muscle fibers in several ways. Cardiac fibers are uni-nucleate, exhibit branching, and are electrochemically coupled to one another by means of intercalated disks containing numerous gap junctions. **(Myocardium)**

11. a) Left sublavian artery

    b) Arch of aota

    c) Pulmonary trunk

    d) Pulmonary semilunar valve

    e) Left atrium

    f) Bicuspid value

    g) Chordae tendinae

    h) Papillary muscle

    i) Left ventricle

    j) Cardiac muscle

    k) Right ventricle

    l) Interventricular septum

    m) Papillary muscle

    n) Chordae Tendinae

    o) Tricuspid valve

    p) Right atrium

    q) Superior vena cava

r) Brachiocephalic trunk

s) Left common carotid artery **(Anatomy of the heart)**

12. The blood returning from the systemic circuit enters the right atrium where it is pumped through the tricuspid valve into the right ventricle. The contraction of the right ventricle propels blood through the right semi-lunar valve into the pulmonary trunk and thus to the pulmonary circuit. Blood returning from the pulmonary circuit enters the left atrium. The contraction of the left atrium causes the blood to flow through the bicuspid valve into the left ventricle. Contraction of the left ventricle propels blood through the left semi-lunar valve into the aorta and thus to the systemic circuit. **(Blood flow through the heart)**

13. The myocardium, like other muscle tissues, has a pronounced requirement for nutrients and oxygen. The endocardium of the heart is impermeable to blood, so a system to deliver oxygenated and nutrient rich blood to the working myocardium is necessary. That system is the coronary circulation system, composed of coronary arteries which branch from the ascending aorta producing capillary nets in the walls of the heart. These capillaries converge to form coronary veins to return the blood from the myocardium. The coronary veins drain into a coronary sinus that empties into the right atrium. **(Coronary circulation)**

14. Autorhythmic cells are self-exciting cells. Approximately 1 percent of cardiac cells are autorhythmic. These cells form the pacemaker and conduction system for the heart. The sinoatrial node is the primary pacemaker of the heart. Excitation from this center of autorhythmic cells spreads across the atria from one cardiac muscle fiber to the next across the intercalated disks. A second center of autorhythmic cells, the atrioventricular node, is activated by the spreading excitation. Descending from the atrioventricular nodes are the bundles of His which course along the interventricular septum to separate into right and left bundles that enter the myocardium of the ventricles and spread the activation there. The bundles of His are required due to the inability of the connective tissues of the fibrous skeleton. These tissues separate the atria from the ventricles and are incapable of conducting a chemoelectric impulse. **(Physiology of cardiac excitation)**

15. Action potentials spread by the autorhythmic cells will cause the voltage-gated Na+ channels in the sarcolemma of cardiac muscle fibers to open. Due to the action of Na+/K+ pumps, the sarcolemma resting membrane potential is close to -90 mV. Na+ will follow its chemical and electrical gradient and flow rapidly into the cardiac muscle fiber. This rapid depolarization triggers voltage-gated Ca++ channels increasing the concentration of Ca++ in the cytosol. Ca++ binds to troponin, allowing actomyosin formation that allows subsequent contractions. In cardiac muscle fibers, repolarization occurs after a prolonged delay. Voltage-gated K+ channels open, causing a diffusion of K+ outside the cell. As Ca++ gates close, the resting membrane potential is reset. **(Physiology of cardiac contraction)**

16. a) P

    b) P-Q interval

    c) Q

    d) R

    e) S

    f) T

g) QRS complex

h) Q-T interval

i) S-T interval **(Electrocardiagram)**

17. c **(Electrocardiogram)**

18. d **(Electrocardiogram)**

19. e **(Electrocardiogram)**

20. b **(Electrocardiogram)**

21. a **(Electrocardiogram)**

22. f **(Electrocardiogram)**

23. All of the events associated with one heartbeat is a single cardiac cycle. In the normal cycle, the two atria contract while the two ventricles relax. The two atria then relax as the two ventricles contract. A phase of contraction is called the systole. A phase of relaxation is called a diastole. Each cardiac cycle contains a systole and a diastole for both the atria and ventricles. At the end of a heartbeat, all four chambers are in diastole. This is called the relaxation phase. During this phase, the relaxation of the chambers of the heart results in their passive filling. When ventricular pressure drops below that of atrial pressure, the AV valves open and the second phase, ventricular filling, begins. Firing of the SA node produces contraction across the atria (systole) forcing blood into the ventricles. At the end of the atrial systole, the AV node excites the ventricular systole and blood is ejected from the ventricles into the systemic and pulmonary circuits. **(Cardiac cycle)**

24. The stroke volume of the heart is affected by three factors:

    a. preload — the amount of stretch on the heart

    b. contractility — the forcefulness of contraction of the myocardium of the ventricles

    c. afterload — the pressure that must be exceeded before blood ejection is accomplished **(Cardiac output)**

25. The Frank-Starling law of the heart states that within limits, the greater the preload or stretching of cardiac muscle, the greater the forcefulness of contraction during the systole. **(Cardiac output)**

26. The cardiovascular center is located in the medulla oblongata of the brain. It receives input from higher centers of the brain, notably the cerebral cortex, limbic system, and hypothalamus. It also receives information from sensory receptors such as proprioceptors that monitor movement, chemoreceptors that monitor blood chemistry, and baroreceptors that monitor blood pressure. This data is processed and appropriate responses are initiated by the cardiovascular system through its output to the sympathetic autonomic nervous system and the cardiac accelerator nerves that connect to the SA node, AV node, and directly to the myocardium. Activity by this system results in increased contractility of myocardium and an increased rate of spontaneous depolarization in the SA and AV nodes. The cardiovascular center also affects the activation of the vagus nerve of the parasympathetic branch of the autonomic nervous system.

Branches of the vagus innervate the SA and AV nodes, as well as directly affecting myocardium. Activity by the vagus nerve decreases the rate of spontaneous depolarization of the SA and AV nodes. **(Regulation of heart rate)**

27. The two chemicals that directly affect the activation of the heart are hormones and ions. The delivery of epinephrine and nor-epinephrine by the endocrine system into the blood stream have a similar effect to the activity of cardiac accelerator nerves. Na^+ and K^+ are critical to the production of all action potentials. Ca^{++} is critical to the contractility of muscles. Imbalances in these electrolytes in the blood greatly affect the ability of the heart to contract. **(Regulation of the heart rate)**

Grade Yourself

Circle the numbers of the questions you missed, then fill in the total incorrect for each topic. If you answered more than three questions incorrectly, you need to focus on that topic. (If a topic has less than three questions and you had at least one wrong, we suggest you study that topic also. Read your textbook, a review book, or ask your teacher for help.)

Subject: The Cardiovascular System: The Heart

Topic	Question Numbers	Number Incorrect
Anatomy of the heart	1, 2, 8, 11	
Chambers of the heart	3, 4	
Heart rate	5	
Valves of the heart	6	
Cardiac output	7	
Pericardium	9	
Myocardium	10	
Blood flow through the heart	12	
Coronary circulation	13	
Physiology of cardiac excitation	14	
Physiology of cardiac contraction	15	
Electrocardiagram	16, 17, 18, 19, 20, 21, 22	
Cardiac cycle	23	
Cardiac output	24, 25	
Regulation of heart rate	26, 27	

The Cardiovascular System: Arteries, Capillaries, and Veins

Brief Yourself

The blood vessels may be described as the conduits of the cardiovascular system. The heart functions as a muscular pump to propel the blood throughout the body and the blood functions as our medium of transport and exchange for our wastes, nutrients, and oxygen. The blood vessels function as the pipelines to carry blood from the heart to the tissues.

Blood vessels are of three types; arteries, capillaries, and veins. Arteries convey blood from both ventricles of the heart to virtually every part of the body. They are found in all tissues with the exceptions of the hair, nails, epidermis, cartilage, and cornea. Arteries branch to become smaller vessels, arterioles, which in turn branch to produce capillaries. Capillaries are the ultimate subdivision of the vascular system and it is through the capillaries that the exchange of carbon dioxide, oxygen, water, nutrients, metabolites, and wastes occur between the blood and tissues. Capillaries converge to form the venules, which continue convergence to form the larger veins. Veins serve to return blood to the heart, either from the systemic circuit or from the pulmonary circuit.

Test Yourself

1. Label the following diagram of an artery, vein, and capillary.

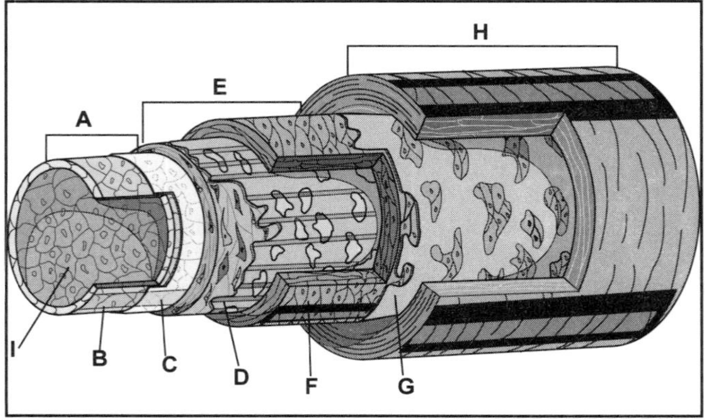

Fig. 14-1

2. What is the difference between elastic (conducting) arteries and muscular (distributing) arteries?

3. What is microcirculation? Trace the flow of blood through the vascular structures from an arteriole to a venule.

4. How do veins differ anatomically from arteries?

5. When your body is at rest how is the blood distributed in the body? Why are veins and venules often called blood reservoirs?

6. What are the three processes that produce capillary exchange?

7. What is Starling's law of capillaries? What forces act to produce blood filtration and reabsorption at the capillaries?

8. What is edema? What are the main causes of edema?

9. What determines the velocity of blood flow? How does it vary throughout the vascular system?

10. How is the volume of blood flow determined? What is an average blood flow in a normal adult?

11. What is blood pressure? Why do the diastolic and systolic pressures vary?

12. How is blood pressure typically measured? What are Korotkoff sounds?

13. What three factors produce the resistance to blood flow?

14. Due to the relatively low pressure of the venous system, what two mechanisms other than the heart are important to venous return?

15. How does the neural system regulate blood pressure?

16. How are hormones active in the regulation of blood pressure?

17. What is autoregulation of blood pressure? What two factors affect autoregulation?

18. Label the following diagram of the major arteries of the thorax and abdomen.

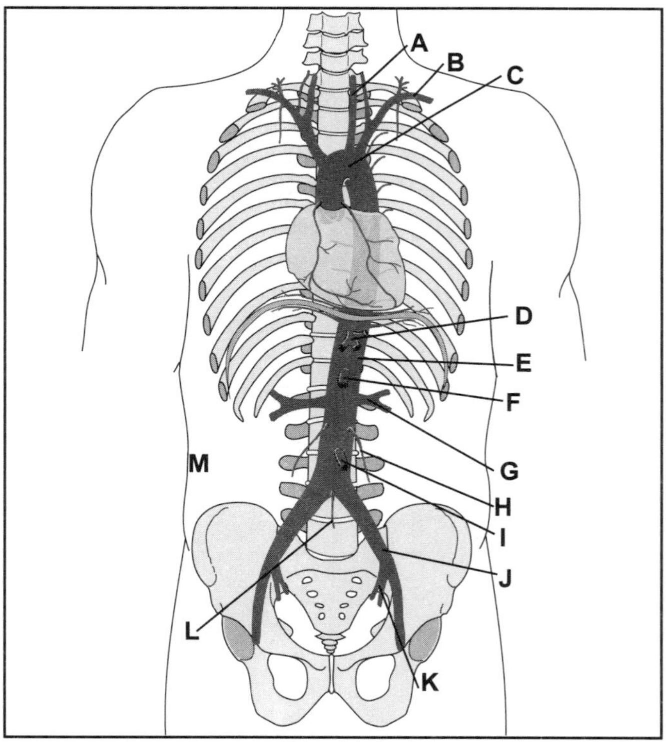

Fig. 14-2

19. Label the following diagram of the major veins of the thorax and abdomen.

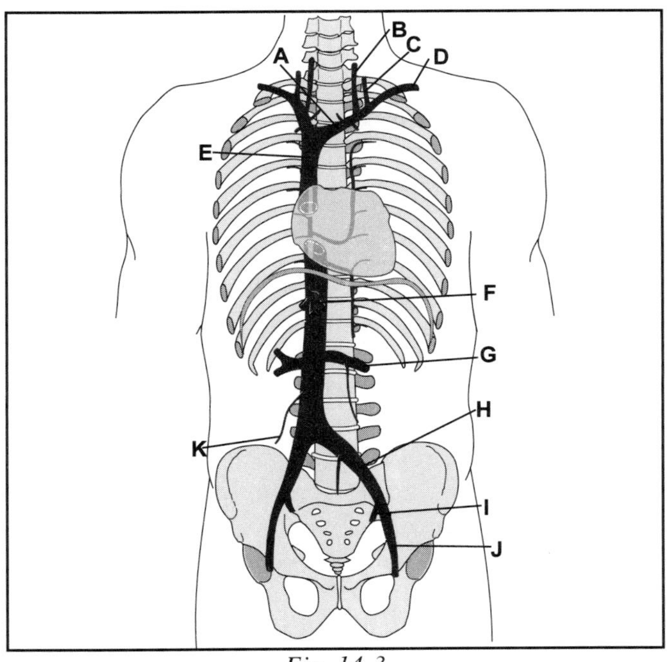

Fig. 14-3

20. Label the following diagram of the major arteries of the head.

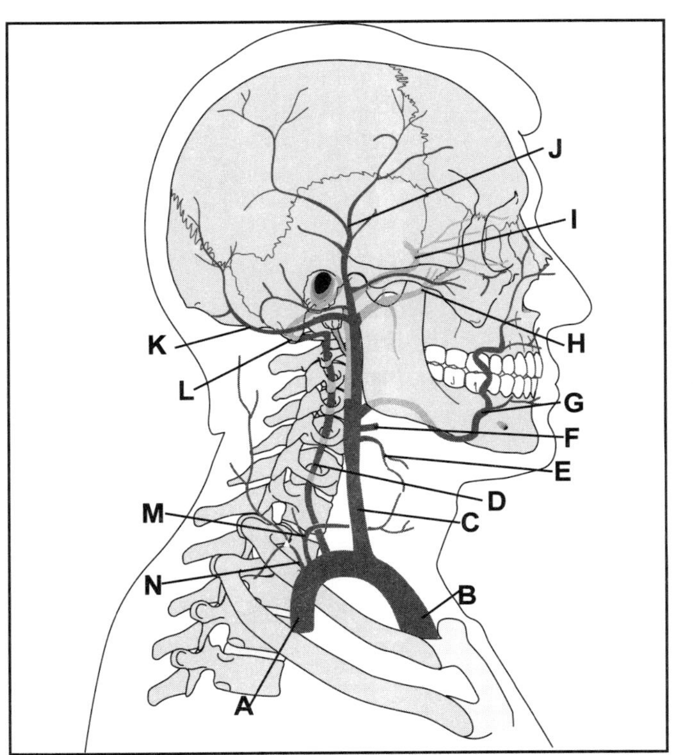

Fig. 14-4

21. Label the following diagram of the major veins of the head.

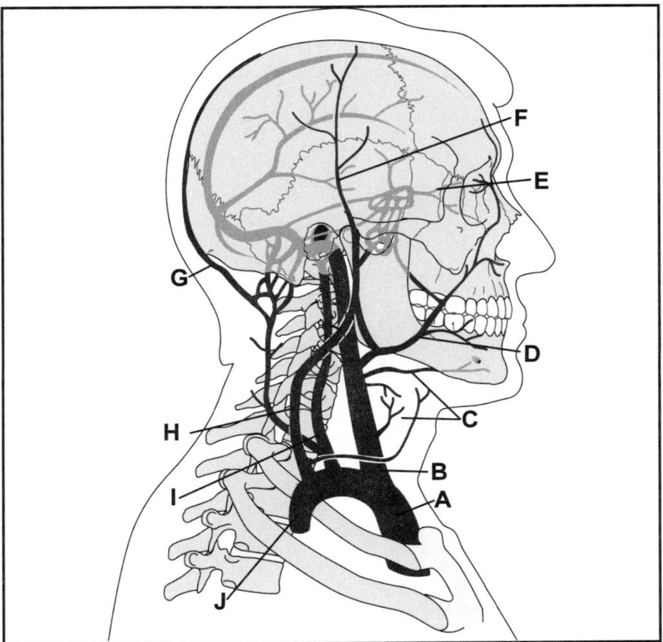

Fig. 14-5

22. Label the following diagram of the major arteries of the upper limb.

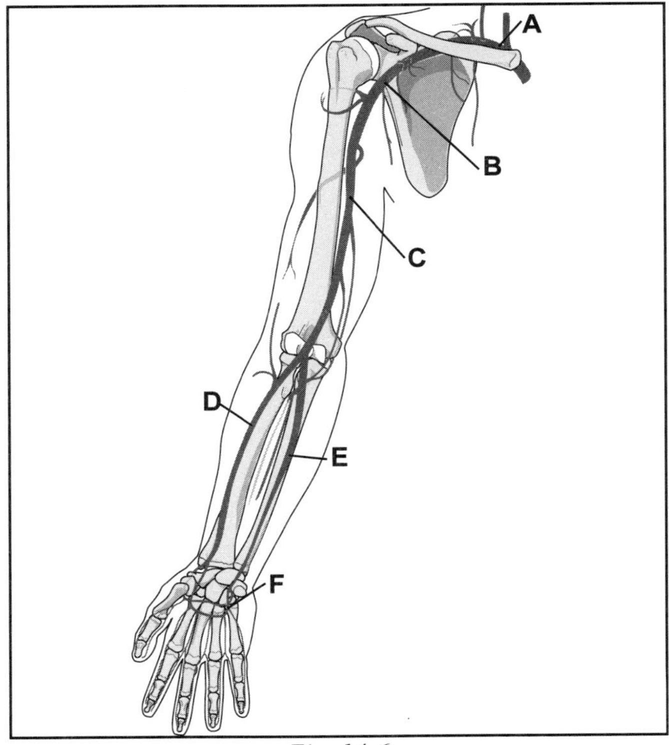

Fig. 14-6

23. Label the following diagram of the major veins of the upper limb.

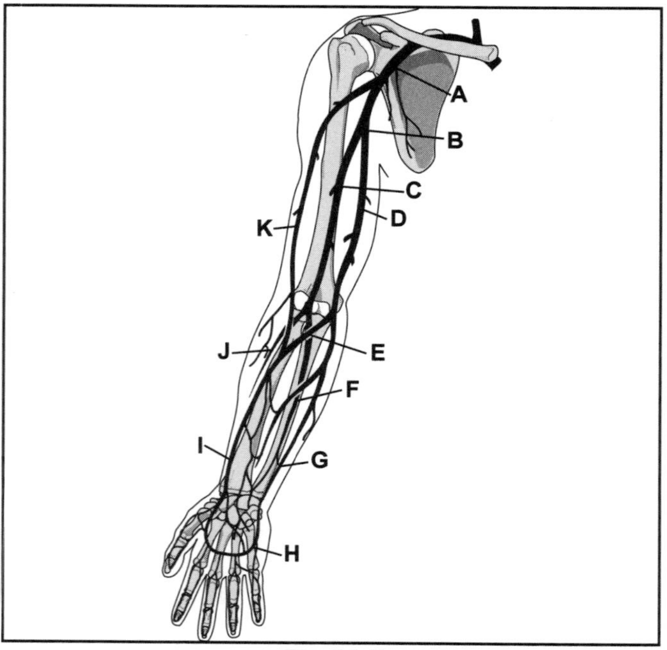

Fig. 14-7

24. Label the following diagram of the major arteries of the lower limb.

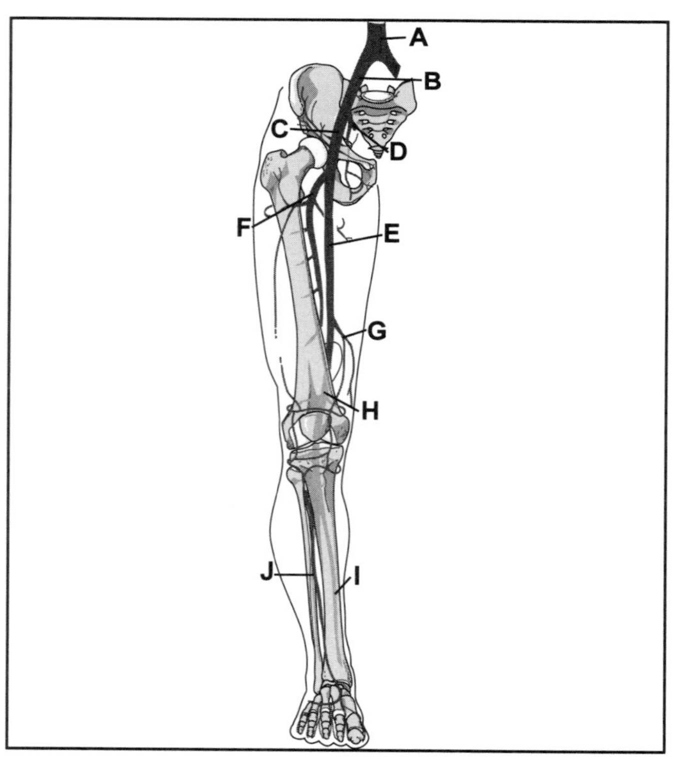

Fig. 14-8

25. Label the following diagram of the major veins of the lower limb.

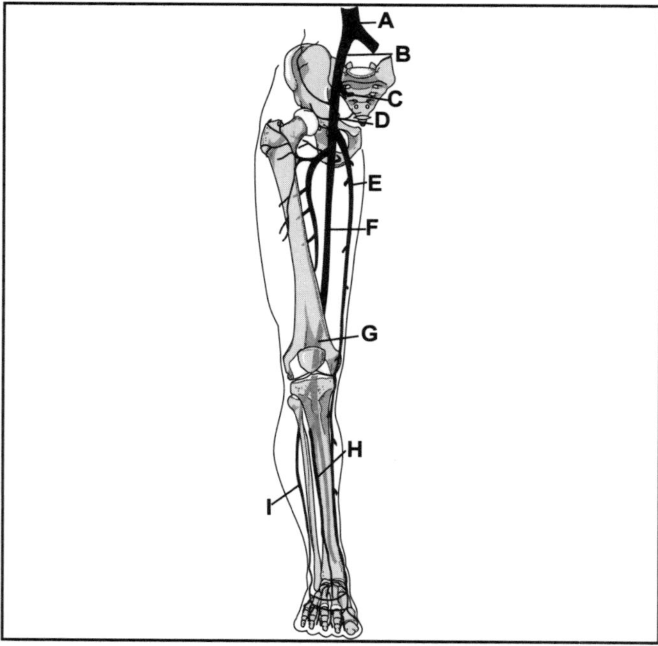

Fig. 14-9

26. What is a portal system? Why are they important?

27. Trace the blood flow through the structures of the pulmonary circuit. What blood pressures might be expected in this system?

Check Yourself

1. a) Tunica interna

   b) Endothelium

   c) Basement membrane

   d) Internal elastic lamina (artery only)

   e) Tunica media

   f) Smooth muscle

   g) External elastic lamina (artery only)

   h) Tunica externa

   i) Lumen **(Anatomy of blood vessels)**

2. The elastic arteries are the larger arteries that have relatively thin walls relative to the size of the lumen. The tunica media of these arteries contain a larger proportion of elastic connective tissue and a lesser proportion of smooth muscle when compared to the muscular arteries. Elastic arteries are often called conducting arteries because they conduct blood from the heart to the muscular arteries. Muscular arteries are often called distributing arteries because these arteries distribute the blood to the various parts of the body. **(Anatomy of blood vessels)**

3. Microcirculation refers to the flow of blood from the arterioles to the venules through the microscopic capillaries. Arterioles branch to form metarterioles that pass through capillary networks to empty into venules. The proximal portion of the metarteriole contains smooth muscle fibers that act as precapillary sphincters to control blood flow into the capillary network. The distal portion of the metarteriole has no smooth muscle tissue and empties into the venule. This portion of the metarteriole is called a thoroughfare channel. True capillaries branch from the metarteriole and are not a part of the direct flow from the metarteriole to the thoroughfare channel. Each true capillary will have a precapillary sphincter at its proximal end to control blood flow and will eventually connect to the thoroughfare channel at its distal end. **(Anatomy of blood vessels)**

4. Veins have essentially the same lamina as arteries except for the elastic lamina. The other layers are similar, but typically veins have a much thinner tunica media with less smooth muscle tissue than is found in arteries. The tunica intima of the vein is slightly thinner than their companion arteries and the tunica externa is thicker. In addition, veins are supplied with one-way valves with flap-like cusps similar to the semi-lunar valves of the heart. These restrict blood flow in one direction only, toward the heart. **(Anatomy of blood vessels)**

5. With the body at rest, the greatest portion of the blood is found in the systemic veins and venules, approximately 60 percent. For this reason, the veins and venules are often called the blood reservoir. At rest, capillaries only contain approximately 5 percent of the blood volume, while the arteries and arterioles hold about 15 percent. **(Blood distribution)**

6. Capillary exchange refers to the substances that enter and exit from the vascular system at the capillary networks. The three forces that produce this exchange are diffusion, vesicular transport, and bulk flow (filtration and reabsorption). **(Capillary exchange)**

7. The Starling law of capillaries states that the volume of blood filtered from the capillaries is slightly greater than the volume reabsorbed. Four basic pressures affect the net filtration pressure (NFP). Two hydrostatic pressures are involved. The first is pressure of the water in the fluid component of the blood that presses against the capillary walls, which is called blood hydrostatic pressure (BHP). The second is the opposing pressure of the water found outside the capillaries in the interstitial fluid. This is called interstitial fluid hydrostatic pressure (IFHP). At the arterial end of the capillaries, the BHP is approximately 35 mmHg. The IFHP is essentially at 0mmHg all along the length of the capillaries. At the venous end of the capillaries, due to the loss of fluid from the capillaries, the BHP drops to about 16 mmHg. Similarly, two osmotic pressures are involved. The presence of plasma proteins in the capillaries produces most of the osmotic pressure because these proteins are too large to escape from the capillaries. The blood colloid osmotic pressure (BCOP) is the force produced by the suspension of these large proteins in the blood remaining in the capillaries. Opposing the BCOP is the interstitial fluid osmotic pressure (IFOP) produced by the solutes in the interstitial fluid. The BCOP remains relatively constant throughout the capillaries at about 26 mmHg. The IFOP varies slightly, but can be approximated at a value of 1 mmHg for most conditions. Net filtration pressure can then be determined as the result of the pressures forcing fluid from the capillaries opposed by the pressures forcing fluid into the capillaries. The following formula will determine the direction of fluid movement:

$$\text{NFP} = (\text{BHP} + \text{IFOP}) - (\text{BCOP} + \text{IFHP})$$

At the arterial end of the capillary, these values are:

$$\text{NFP} = (35 + 1) - (26 + 0) = 17 - 26 = -9\text{mm Hg for outward flow}$$

At the venous end of the capillary, these values are:

$$\text{NFP} = (16 + 1) - (26 + 0) = 17 - 26 = -9\text{mm Hg for inward flow}$$

This indicates that more fluid is filtered from the capillaries than is reabsorbed. Typically, about 85 percent of the fluid filtered from the blood is reabsorbed at the venous end of the capillaries. The remainder of the fluid is absorbed into the lymphatic system. **(Capillary exchange)**

8. Edema is the accumulation of fluid in the tissue (interstitial fluid) that occurs when filtration greatly exceeds reabsorption. The most common causes of edema are:

   a. increased hydrostatic pressure in the capillaries.

   b. decreased concentration of plasma proteins.

   c. increased permeability of capillaries.

   d. increased extracellular fluid volume.

   e. blockage of lymphatic vessels. **(Capillary exchange)**

9. The velocity of blood flow is inversely related to the cross-sectional area of the blood vessels through which it travels. As the cross-sectional area increases, the velocity of blood flow will decrease. Considering the cross-sectional area of the entire vascular system, blood velocity will be greatest as it exits the heart into the arterial system. As the arterial system continues to branch, increasing its cross-sectional area, blood velocity will decrease reaching its slowest speed through the capillaries. The

velocity will increase as the blood enters the venules as they converge producing a decreasing cross-sectional area. **(Hemodynamics)**

10. The volume of blood flow, or cardiac output (CO), is a consequence of the stroke volume (SV) of the heart and the heart rate (HR), such that:

$$CO = SV \times HR$$

The average stroke volume is about 70 ml/beat and the average heart rate is about 75 beats/min, thus, the average CO can be determined:

$$CO = 70 \text{ ml/min} \times 75 \text{ beats/min} = 5250 \text{ ml/min}$$

(Hemodynamics)

11. Blood pressure is the hydrostatic pressure exerted by the fluid of the blood as it presses against the blood vessel walls. During the contraction (systole) of the heart, the propulsion of blood into the arterial system produces a rise in pressure. This systolic pressure is about 120 mmHg for the average healthy adult. When the heart is in relaxation (diastole), the pressure in the arterial system drops to about 80 mmHg. **(Hemodynamics)**

12. Blood pressure is typically measured at left brachial artery using a sphygmomanometer. The cuff of the sphygmomanometer is secured around the arm and inflated until the cuff exceeds the pressure in the artery, restricting blood flow. As the cuff is slowly deflated, a stethoscope is utilized to listen to the sounds emanating from the brachial artery. The Korotkoff sounds indicate activity in the artery. The first Korotkoff sound is a faint "tapping" sound that indicates that the pressure of the cuff is equal to the pressure in the artery as blood flow resumes with the systole. As the pressure of the cuff continues to drop, the sound will become faint as the turbulence drops. At this point, the diastolic pressure may be read. **(Measurement of blood pressure)**

13. Resistance to blood flow refers to the opposition to blood flow that is primarily a function of resistance between the blood and the walls of the vessels. The three factors producing resistance to blood flow are blood viscosity, total length of the blood vessels, and the average blood vessel radius. **(Hemodynamics)**

14. In addition to the action of the heart, the two forces that support venous return are the skeletal muscle pump and the respiratory pump. Because the walls of veins are less muscular than arteries, they are more responsive to pressure produced around them. The contractions of skeletal muscles tend to collapse the veins, pressing the blood against the one-way valves of the vein, milking the blood toward the heart. During inspiration, the inferior movement of the diaphragm produces a decrease in thoracic pressure and an increase in abdominal pressure. This produces a movement of the blood in the compressed abdominal veins into the thoracic area. **(Hemodynamics)**

15. Neural regulation of the cardiovascular system depends upon sensory receptor information from baroreceptors in the walls of the arteries and right atrium, and chemoreceptors located in the carotid and aortic bodies. These sensory receptors provide information to the cardiovascular center (CV) of the medulla oblongata. Increases in blood pressure or increases in hydrogen ions or carbon dioxide content of the blood causes the CV to increase the parasympathetic activation to the heart and blood vessels. This will result in lowering the heart rate, lowering the force of cardiac contraction, and increasing vasodilation. A similar effect occurs if oxygen levels in the blood fall. If the blood pressure decreases, the decrease in activation of the CV results in the increase of sympathetic activity causing an increased heart rate, increased contractility of the myocardium, and increased vasoconstriction. **(Regulation of blood pressure)**

16. Then endocrine system regulates the blood pressure primarily by controlling fluid retention in the body (affecting blood volume), contributing to setting heart rate and influencing vasodilation or vasoconstriction. The following hormones affect blood pressure:

a. renin-angiotensin-aldosterone system.

b. epinephrine and nor-epinephrine.

c. antiduretic hormone.

d. atrial natriuretic peptide.

e. parathyroid hormone and calcitrol. **(Regulation of blood pressure)**

17. Autoregulation of blood pressure refers to the effect produced by local, automatic adjustments to blood flow in a given area of the body. Physical changes, like temperature, produce vasodilations in response to increased temperature or vasoconstrictions in response to decreased temperatures. The smooth muscles of blood vessels also exhibit a myogenic response to stretch, contracting more forcefully when stretched and less so when relaxed. A variety of chemicals produce local responses in the blood vessels. Among these are vasodilators such as nitric oxide, lactic acid, and adenosine. Vasoconstrictors include eicosanoids, angiotensin, and endothelins. **(Regulation of blood pressure)**

18. a) Left common carotid

b) Left subclavian

c) Aortic arch

d) Celiac trunk

e) Left adrenal

f) Splenic

g) Left rental

h) Left gonadal

i) Inferior mesenteric

j) Left common iliac

k) Left internal iliac

l) Median sacral

m) Abdominal aorta **(Anatomy of blood vessels)**

19. a) Left brachiocephalic

b) Left internal jugular

c) Left external jugular

d) Left subclavian

e) Superior vena cava

f) Hepatic

g) Left rental

h) Left common iliac

i) Left internal iliac

j) Left external iliac

k) Right gonadal (**Anatomy of blood vessels**)

20. a) Axillary

b) Brachocephalic

c) Common carotid

d) Vertebral

e) Superior thyroid

f) Lingual

g) Facial

h) Maxillary

i) Opthalamic

j) Superior temporal

k) Occipital

l) Basilar

m) Thyrocervical

n) Cosacervical (**Anatomy of blood vessels**)

21. a) Brachiocephalic

b) Internal jugular

c) Thyroids

d) Facial

e) Opthalamic

f) Superior temporal

g) Occipital

h) External jugular

i) Vertebral

j) Subclavia **(Anatomy of blood vessels)**

22. a) Right subclavian

b) Right axillary

c) Right brachial

d) Right radial

e) Right ulnar

f) Right superior palmar arch **(Anatomy of blood vessels)**

23. a) Right subclavian

b) Right axillary

c) Right Brachial

d) Right basalic

e) Right median cubital

f) Right median antebrachial

g) Right ulnar

h) Right palmar arch

i) Right cephalic

j) Right accessory cephalic

k) Right cephalic **(Anatomy of blood vessels)**

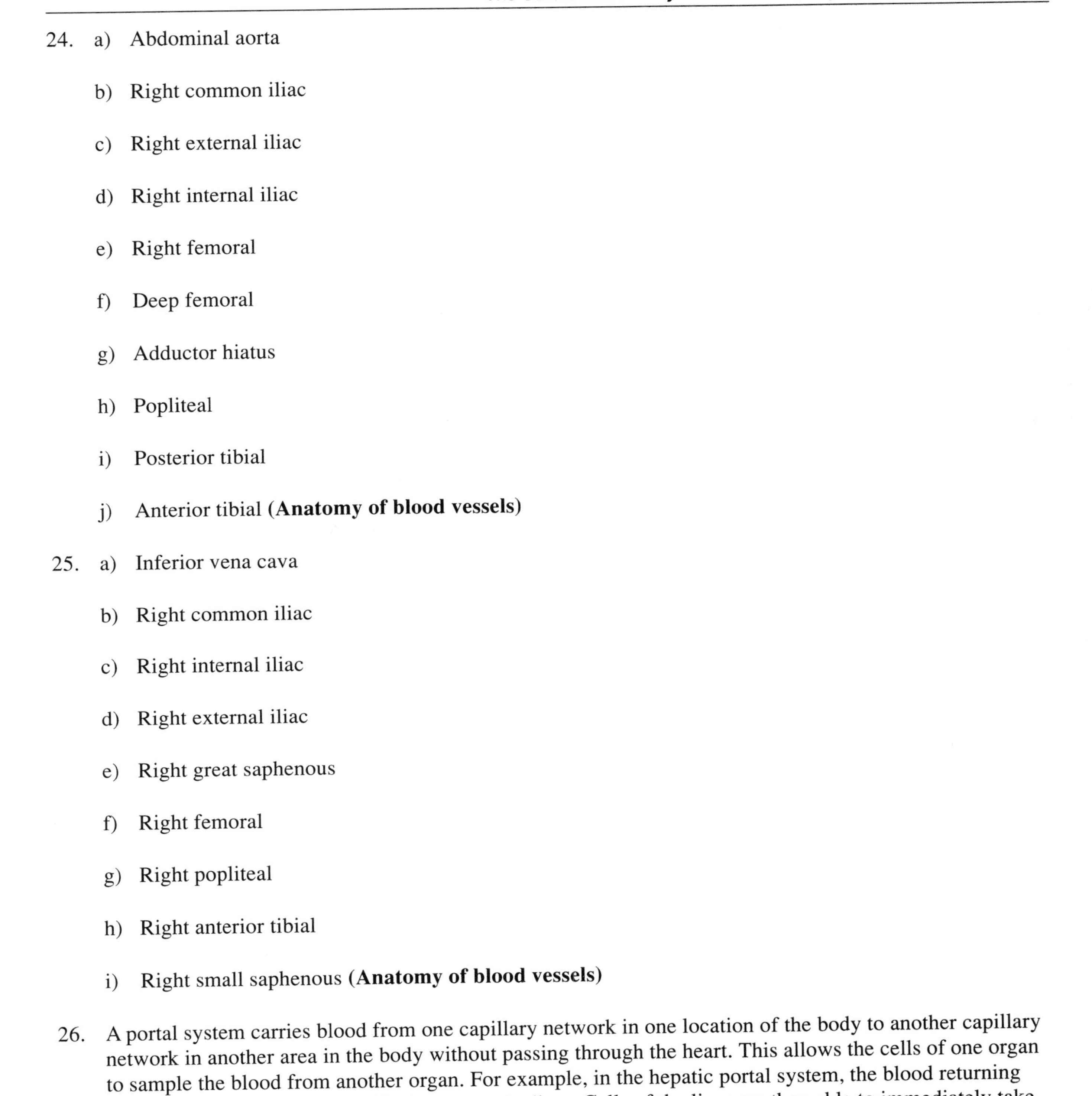

24. a) Abdominal aorta

    b) Right common iliac

    c) Right external iliac

    d) Right internal iliac

    e) Right femoral

    f) Deep femoral

    g) Adductor hiatus

    h) Popliteal

    i) Posterior tibial

    j) Anterior tibial **(Anatomy of blood vessels)**

25. a) Inferior vena cava

    b) Right common iliac

    c) Right internal iliac

    d) Right external iliac

    e) Right great saphenous

    f) Right femoral

    g) Right popliteal

    h) Right anterior tibial

    i) Right small saphenous **(Anatomy of blood vessels)**

26. A portal system carries blood from one capillary network in one location of the body to another capillary network in another area in the body without passing through the heart. This allows the cells of one organ to sample the blood from another organ. For example, in the hepatic portal system, the blood returning from the digestive system capillaries enters the liver. Cells of the liver are then able to immediately take the nutrient rich blood and store or modify its nutrient content. **(Portal circulation)**

27. The pulmonary circuit carries deoxygenated blood returning to the right atrium and then pumped by the right ventricle to the alveoli of the lungs and returns to the left atrium. The right ventricle propels blood into the pulmonary trunk with branches to produce the left and right pulmonary arteries. The continued branching of these arteries eventually produces the capillary networks that surround the alveoli. The pulmonary capillaries converge to form pulmonary veins. There are two left and two right pulmonary veins that enter the left atrium. The blood pressure in the pulmonary circuit is lower than the systemic circuit due largely to the rapid branching of the pulmonary arteries, rapidly increasing the cross-sectional area of the pulmonary vascular system. **(Pulmonary circulation)**

Grade Yourself

Circle the numbers of the questions you missed, then fill in the total incorrect for each topic. If you answered more than three questions incorrectly, you need to focus on that topic. (If a topic has less than three questions and you had at least one wrong, we suggest you study that topic also. Read your textbook, a review book, or ask your teacher for help.)

Subject: The Cardiovascular System: Arteries, Capillaries, and Veins

Topic	Question Numbers	Number Incorrect
Anatomy of blood vessels	1, 2, 3, 4, 18, 19, 20, 21, 22, 23, 24, 25	
Blood distribution	5	
Capillary exchange	6, 7, 8	
Hemodynamics	9, 10, 11, 13, 14	
Measurement of blood pressure	12	
Regulation of blood pressure	15, 16, 17	
Portal circulation	26	
Pulmonary circulation	27	

The Lymphatic System and Immunity

Brief Yourself

The lymphatic system consists of a fluid called lymph, the lymphatic vessels contain it, several structures contain lymphatic tissue, several organs, and the stem cells of the red bone marrow that produce lymphocytes. The lymph is essentially the same as interstitial fluid, but flowing within lymphatic vessels. Lymphatic tissue is considered a specialized form of reticular connective tissue that contains large numbers of lymphocytes.

Our environment presents numerous dangers to our health, including pathogens and tissue trauma. Our ability to neutralize the activity of pathogens and toxins produced by them, as well as the ability to repair tissue damage due to trauma is critical to our homeostasis. The ability to ward off disease is called resistance. Resistance can be grouped into two areas; nonspecific resistance and immunity.

Non-specific resistance refers to the defense mechanisms that provide general protection against a wide variety of pathogens, including bacteria and viruses. Such characteristics as the chemical barriers presented by the skin and mucous membranes, antimicrobial chemicals, nonspecific phagocytes, inflammation, and fever all contribute to our non-specific resistance.

Immunity is a system of cooperating cells and chemical agents that act to defend against specific pathogens or foreign matter. The body system most responsible for our immunity is the lymphatic system.

Test Yourself

1. What are the three major functions of the lymphatic system?

2. Label the following diagram of the major structures of the lymphatic system.

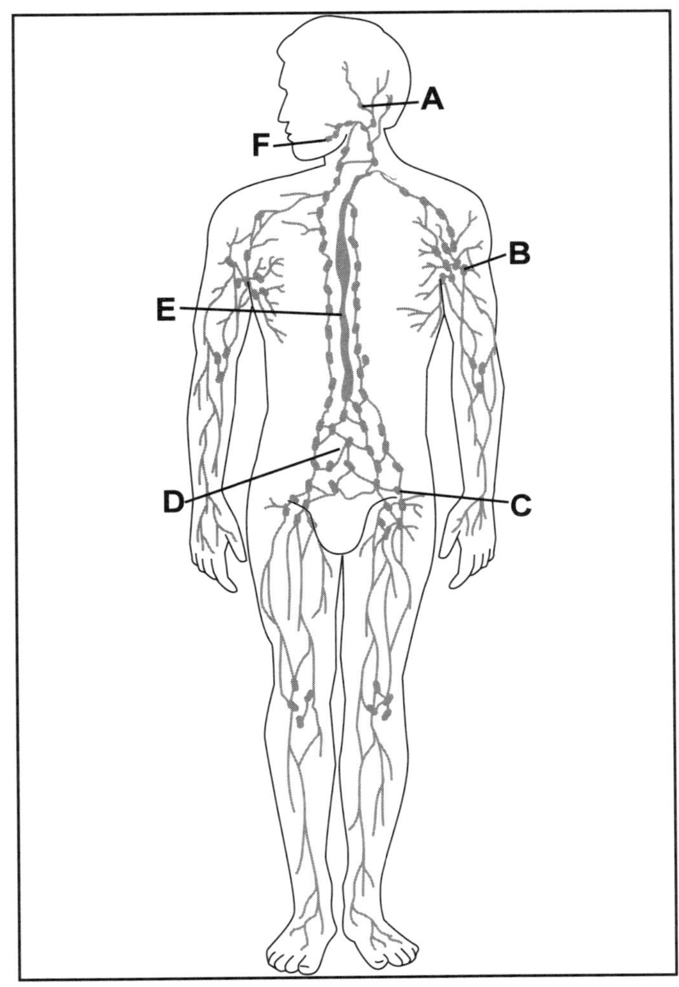

Fig 15-1

3. Where are the lymph nodes located that drain the superficial areas of the head and face?

4. What organs, nodes, and ducts are considered a part of the lymphatic system?

5. Label the following diagram of a lymph node.

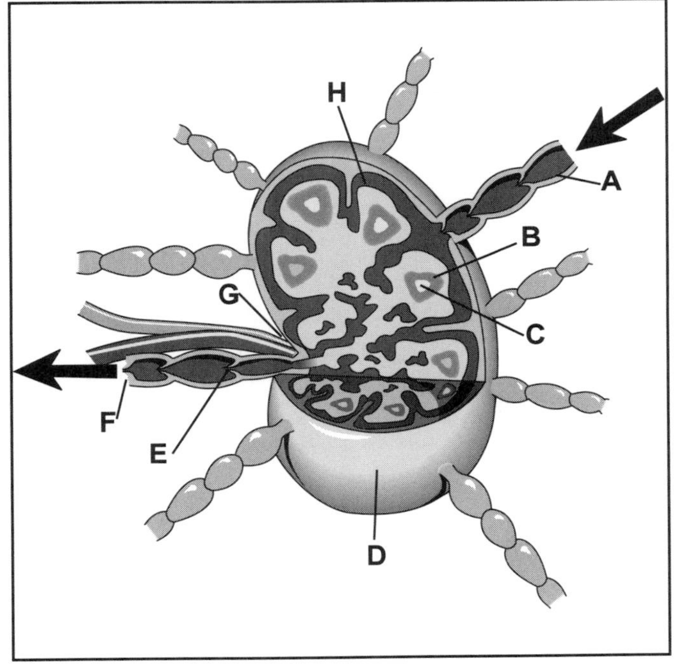

Fig 15-2

6. How is lymphatic fluid produced and collected into the lymphatic capillaries?

7. Where is the thymus gland located? What is its primary function?

8. Trace the flow of lymph through a lymph node. Where are B and T cells found in the lymph node?

9. Where is the spleen located? What are the functions of the spleen?

10. Describe the characteristics of mucous membranes and the skin that produce a mechanical protection against pathogen and foreign substance entry into the body.

11. What chemical protection is provided by the skin and mucous membranes?

12. What is the role of interferon in nonspecific resistance to disease?

13. What is the complement system of the blood? How is complement activated?

14. Once activated, how does complement attack pathogens?

15. What are natural killer cells? How do they differ from T cells?

16. What are the two major types of nonspecific phagocytes? What three phases occur in phagocytosis?

17. What is the role of inflammation in nonspecific resistance? What chemicals are most responsible for the production of inflammation?

18. What two properties distinguish immunity from nonspecific resistance?

19. What is immunocompetence? What is an antigen?

20. What are the two types of immune responses? What are the specializations of each?

21. What is the difference between a complete and partial antigen? What is an epitope?

22. What is the major histocompatibility complex (MHC)?

23. What is antigen-presenting? Trace the steps in antigen-presenting following the phagocytosis of a pathogen by a macrophage.

24. Identify the four basic types of T cells and their function.

Questions 25–29 are matching. Match the following immunoglobulin (Igs) with its characteristics.

25. IgG a. Occurs as dimers or monomers, found primarily in sweat, tears, saliva, milk, mucous membranes, and gastrointestinal secretions

26. IgA b. Occurs as monomers, found on mast cells and basophils

27. IgM c. Occurs as monomers, most common Ig of the blood and lymph

28. IgD d. Occurs as pentamers, found in blood and lymph

29. IgE e. Occurs as monomers, found in blood, lymph, and on the surface of B cells

30. What produces the activation and proliferation of B cells? What two types of B cells are produced?

31. What are the common functions of antibodies?

32. What is the immunological response to cancer?

33. What is the difference between a primary and secondary immunological response?

Check Yourself

1. The lymphatic system drains the interstitial fluid, transports dietary lipids, and carries out immune functions. **(Lymphatic system function)**

2. a) Palatine tonsils

   b) Axillary nodes

   c) Inguinal nodes

   d) Intestinal nodes

   e) Thoracic duct

   f) Submandibular nodes **(Anatomy of lymphatic system)**

3. Lymph nodes draining the superficial areas of the head and face are located slightly inferior to the ear, both anteriorly and posteriorly, and inferior to the mandible. **(Anatomy of lymphatic system)**

4. The primary lymphatic organs are the red bone marrow and the thymus gland. Secondary lymphatic organs include the lymph nodes and spleen. Lymphatic nodules, which are not surrounded by a capsule and thus not considered a discrete organ, are also typically considered a part of the secondary lymphatic organs. The two major collecting ducts of the lymphatic system are the thoracic duct and the right lymphatic duct. **(Anatomy of lymphatic system)**

5. a) Afferent lymphatic vessel

   b) Follicle

   c) Germinal center

   d) Capsule

   e) Valve

   f) Efferent lymphatic vessel

   g) Hilus

   h) Reticular fibers (trabeculae) **(Anatomy of lymphatic system)**

6. Lymphatic fluid is produced as excess blood plasma that is not returned into the capillary system. Approximately 3 liters drain into the lymphatic capillaries per day. Lymphatic capillaries form blind ends in the tissue with large fenestrations. Lymphatic vessels contain one-way valves, similar to those of veins. Like veins, the thin-walled lymphatic vessels are easily collapsed by local skeletal muscle contractions, milking fluid toward the lymphatic ducts and ultimately back to the subclavian veins. **(Lymphatic fluid)**

7. The thymus gland is located in the mediastinum, just superior to the heart and inferior to the thyroid gland. Pre T cells migrate from the red bone marrow to the thymus gland where they accumulate in the medulla. These cells proliferate and develop into mature T cells in the thymus. The thymus also produces thymic hormones thought to be involved in T cell maturation. **(Thymus gland)**

8. Lymph flows through a lymph node in one direction, entering through an afferent lymphatic vessel to flow into the sinuses. Lymph flows through the sinus, passing first through the cortex and then the medulla, to exit through an efferent lymphatic vessel. T cells are found aggregated in the outer layers of the follicles of the cortex. B cells are clustered in the germinal centers deep in each follicle. **(Lymph nodes)**

9. The spleen is located between the stomach and the diaphragm in the left hypochondriac region. The spleen does not filter lymph, but does contain masses of white pulp lymphatic tissue composed of primarily B cells. Its immune function is to provide a site of B cell's proliferation. The principal function of the spleen is the phagocytosis of bacteria and damaged erythrocytes in the red pulp tissue. **(Spleen)**

10. The epidermis of the skin is composed of many layers of densely packed, keratinized cells that provide little space for microbial invasion. Shedding the outer most layer also helps to remove microbes from the skin surface. Mucous membranes secrete mucus, a slightly viscous fluid that will trap many microbes and foreign particles. The mucous membrane of the nose contains mucus-coated hairs that add additional filtering. The mucous membrane of the upper respiratory tract includes ciliated cells that propel mucus to the throat for swallowing. Microbes and foreign particles trapped in the mucus are carried along to be delivered to the stomach. **(Nonspecific resistance)**

11. Sebaceous glands of the skin secrete sebum that forms a protective film on the surface of the skin. The acidity of sebum inhibits the growth of many pathogenic bacteria. Perspiration not only helps flush microbes from the surface of the skin, but also contains lysozyme which can break down the cell walls of many bacteria. Both mucus and perspiration contain immunoglobulins. **(Nonspecific resistance)**

12. Interferons are proteins produced and released by cells invaded by viruses. Interferon diffuses to neighboring cells and binds to surface receptors. Interferon activates the uninfected cells to produce antiviral proteins that can inhibit viral infection. **(Nonspecific resistance)**

13. Complement is a group of about 20 plasma proteins that enhance immune or inflammatory reactions. Complement may be activated by the classical pathway or the alternative pathway. A plasma protein termed C3 will respond to antigen-antibody complexes to begin a cascade effect in the complement proteins in the classical pathway. C3 can also respond to certain polysaccharides on the surface of microbes in the alternative pathway. **(Nonspecific resistance)**

14. Once activated, complement enhances inflammation by acting as a vasodilator, causing mast cells to release histamine, and attacking phagocytes to the area of infection. In addition, complement produces opsonization to promote phagocytosis. Complement can also kill microbes directly by forming a membrane attack complex (MAC) that produces cytolysis. **(Nonspecific resistance)**

15. Natural killer cells have the ability to attack a variety of pathogens unlike T cells which are specialized for specific pathogens. Natural killer cells do not mature in the thymus and do not have antigen receptors in their plasma membranes. **(Nonspecific resistance)**

16. The two major phagocytes are neutrophils and macrophages. Both are relatively generalized phagocytes that can attack a variety of pathogens (neutrophils work best against bacteria). Macrophages develop from monocytes. Phagocyte activity begins with chemotaxis, the attraction of phagocytes by chemical messages, followed by adherence. Adherence occurs when the phagocyte attaches its plasma membrane to the microbe or foreign material to be ingested. Following adherence, ingestion begins by endocytosis. **(Nonspecific resistance)**

17. Inflammation produces redness, swelling, heat, and pain at the site of infection or injury. The swelling and redness are produced by vasodilation at the site. The increased pressure produced by the edema following vasodilation is responsible for the pain and the heat is due to a respiratory burst of activity of phagocytes and repair cells in the area. The principal chemicals responsible for inflammation are histamine, kinins, leukotrienes, complement, and prostglandins. **(Nonspecific resistance)**

18. Immunity demonstrates two properties that are absent in nonspecific resistance; specificity and memory. Specificity is due to responses by the immune system to particular antigens and memory allows for a more rapid and virulent response to a second encounter with the same pathogen. **(Immunity)**

19. Immunocompetence is the maturation of B and T cells, such that they can carry out immune responses if properly stimulated. The B and T cells develop distinctive surface proteins, some of which serve as antigen receptors. Antigens are large, complex molecules, typically proteins, that the immune system will recognize as foreign to the body. They have two characteristics; they will provoke an immune response (immunogenicity) and they will react with produced antibodies (reactivity). **(Immunity)**

20. Immune responses are cell-mediated or antibody-mediated (humoral). Activated T cells produce the cell-mediated response, proliferating into several classes of T cells, including the cytotoxic T cell which actively attacks antigens. The cell-mediated response is effective against intracellular pathogens, some cancer cells, and foreign tissue transplants. The antibody-mediated response is produced by activated B cells that proliferate to produce antibodies. Helper T cells enhance the activation of both the cell-mediated and antibody-mediated responses. Antibody-mediated responses work well against antigens dissolved in the body fluids and extracellular antigens. **(Immunity)**

21. Complete antigens are chemicals that produce immunogenicity and reactivity. Partial antigens (haptens) produce reactivity but not immunogenicity. Many allergic reactions are due to partial antigens. An epitope is the specific portion of an antigen that triggers the immune response. **(Immunity)**

22. The major histocompatibility complex (MHC) is a group of glycoproteins unique to each individual. The MHC is built into the plasma membrane of all the body cells except the erythrocytes. A specialized MHC is found on the surface of antigen-presenting cells, such as T cells. The immune system recognizes the presence of the MHC as an identifier of "self" so that the cells of your body are not attacked by your immune system. **(Immunity)**

23. Although B cells can respond to antigens found in the fluids of the body, T cells must be presented with an antigen associated to its MHC. Antigen presenting occurs when a phagocytic cell ingests an antigen and then displays it on its plasma membrane along with the MHC that phagocytic cells will have on its membrane. This presentation can act to trigger the T cell response. After a macrophage ingests an antigen, the antigen is broken into peptide fragments by digestive enzymes. At the same time, the macrophage produces MHC on its endoplasmic reticulum and packages it into vesicles at the Golgi apparatus. Vesicles containing MHC and the peptide fragments of the antigen fuse, allowing the MHC and peptides to bond. The vesicles carry the MHC-antigen complex to the plasma membrane where they are presented. **(Immunity)**

24. The four types of T cells are:

    a. cytotoxic T cell — causes death of foreign cells by releasing perforin and lymphotoxin; releases cytokines to attract macrophages.

    b. helper T cell — cooperates with B cells to enhance antibody production and secretes interleukin II which stimulates B and T cell proliferation.

    c. suppressor T cell — regulates immune response by producing cytokines that inhibit the proliferation of T cells.

    d. memory T cell — remains in lymphoid tissue to enhance secondary response. **(Immunity)**

25. c **(Immunity)**

26. a **(Immunity)**

27. d **(Immunity)**

28. e **(Immunity)**

29. b **(Immunity)**

30. B cells typically remain in the lymphatic system and become activated when they come in contact with their antigen. B cells have the capability of producing and presenting the MHC-antigen complex. Helper T cells that recognize the MHC-antigen complex provide costimulation to enhance the proliferation and specialization of B cells. Activated B cells produce a clone of plasma B cells that rapidly produce and release antibodies. Activated B cells that do not become plasma B cells remain in the lymphoid tissue as memory B cells to provide an enhanced secondary response. **(Immunity)**

31. Although the different classes of antibodies (immunoglobulins) have certain specialization, their common actions include:

    a. neutralizing antigen.

    b. immobilization of bacteria.

    c. agglutination and precipitation of antigen.

    d. activation of complement.

    e. enhancement of phagocytosis.

    f. provision of fetal immunity. **(Immunity)**

32. When a cell becomes cancerous, it often will display tumor antigens, molecules that are rarely displayed on the surface of normal cells. The immune system can often recognize the tumor antigens as "nonself" and produce a response. T cells, macrophages, and natural killer cells are the most common agents to resist cancer. **(Immunity)**

33. One of the cardinal characteristics of the immune system is its memory for antigens that have triggered an immune response previously. The primary response occurs when the immune system recognizes an antigen for the first time. At the initial contact, there will only be a few immune cells specialized to respond, so the virulence of the response is delayed until these immune cells can divide and produce large numbers of cells capable of producing resistance. At the second contact, large numbers of memory cells (memory T and memory B cells) will be found in the body, remaining from the initial response. With larger numbers of responding cells available, the speed and efficiency of the immune response is greatly elevated. **(Immunity)**

Grade Yourself

Circle the numbers of the questions you missed, then fill in the total incorrect for each topic. If you answered more than three questions incorrectly, you need to focus on that topic. (If a topic has less than three questions and you had at least one wrong, we suggest you study that topic also. Read your textbook, a review book, or ask your teacher for help.)

Subject: The Lymphatic System and Immunity

Topic	Question Numbers	Number Incorrect
Lymphatic system function	1	
Anatomy of lymphatic system	2, 3, 4, 5	
Lymphatic fluid	6	
Thymus gland	7	
Lymph nodes	8	
Spleen	9	
Nonspecific resistance	10, 11, 12, 13, 14, 15, 16, 17	
Immunity	18, 19, 20, 21, 22, 23, 24, 25, 26, 27, 28, 29, 30, 31, 32, 33	

The Respiratory System

16

Brief Yourself

Respiration is the exchange of gases between the atmosphere and tissues of the body. Within our cells, metabolic reactions that release the energy in our nutrients to produce ATP occur continuously. Oxygen is utilized in this process and carbon dioxide is released. Increased carbon dioxide concentration in the blood produces increased acidity that can accumulate in toxic levels. For this reason, a constant source of oxygen is required in the tissues, as well as a constant elimination of carbon dioxide. The respiratory system is responsible for gas exchange, oxygen uptake, and elimination of carbon dioxide. The cardiovascular system is responsible for the transport of gases in the blood from the lungs to the tissues. The respiratory system also contains receptors for our sense of smell and structures for our production of vocalizations.

Three stages of respiration are responsible for gas exchange. Pulmonary ventilation is breathing, the inspiration and expiration of air between the lungs and the atmosphere. External respiration is the exchange of gases that occurs between the alveolar spaces and the blood in pulmonary capillaries. Internal respiration is the exchange of gases between the blood in systemic capillaries and the tissues of the body. Cellular respiration describes the metabolic reactions within cells that consume oxygen and release carbon dioxide. These reactions will be considered in Metabolism and Nutrition (Chapter 18).

Test Yourself

1. Label the following diagram of a sagittal section of the head and neck.

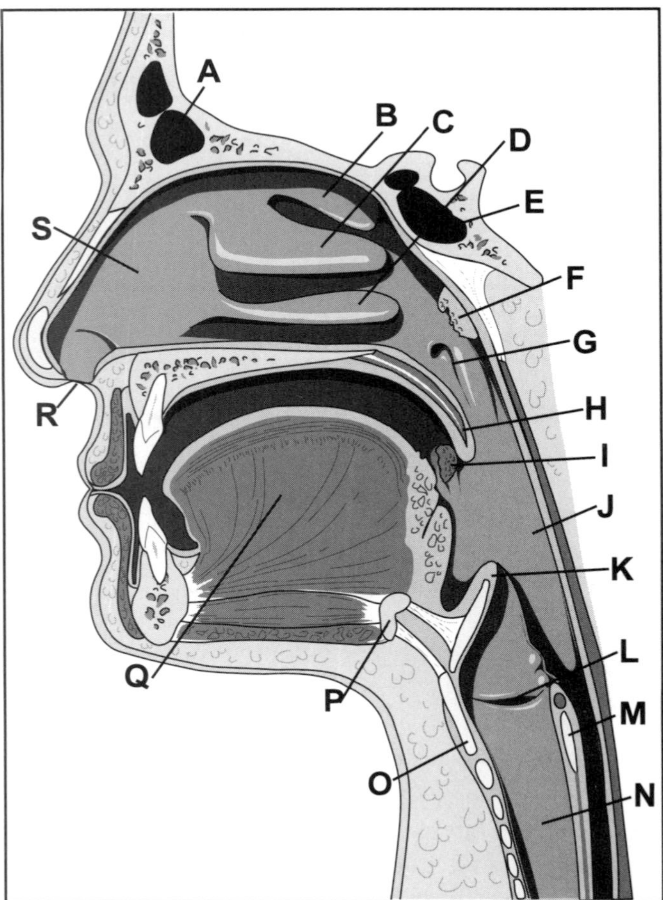

Fig. 16-1

2. What are the functions of the internal structures of the nose?

3. Describe the mucus membrane that lies inferior to the olfactory epithelium. What are its characteristics and functions?

4. Label the following diagram of a sagittal section of the larynx.

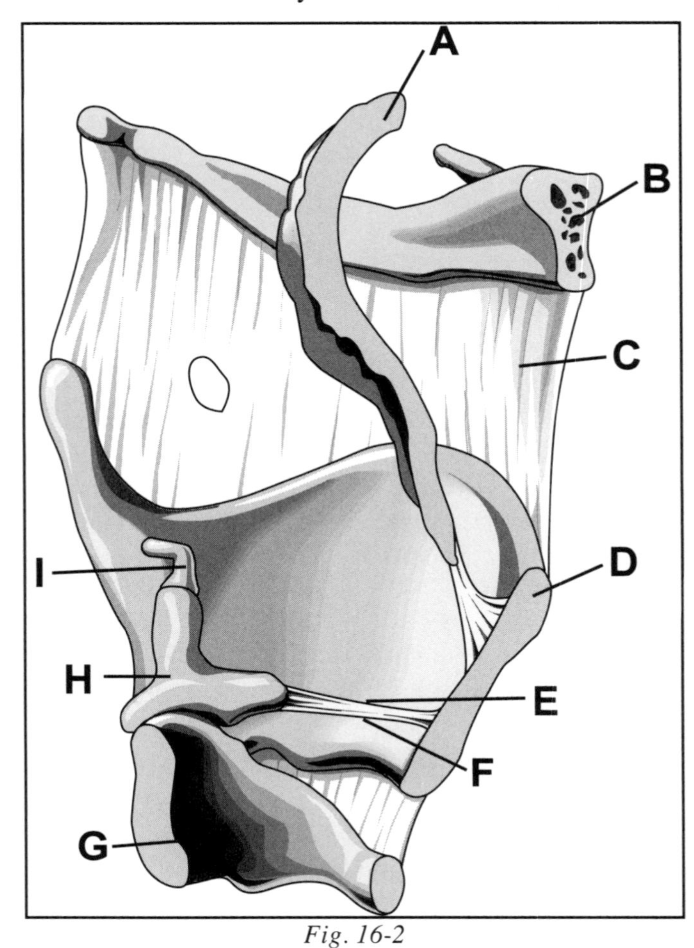

Fig. 16-2

5. Describe the epiglottis and its function.

6. How are vocalizations produced? How are different pitches or intensities of sound produced?

7. What is the role of the tracheal cartilage?

8. Label the following diagram of the lungs.

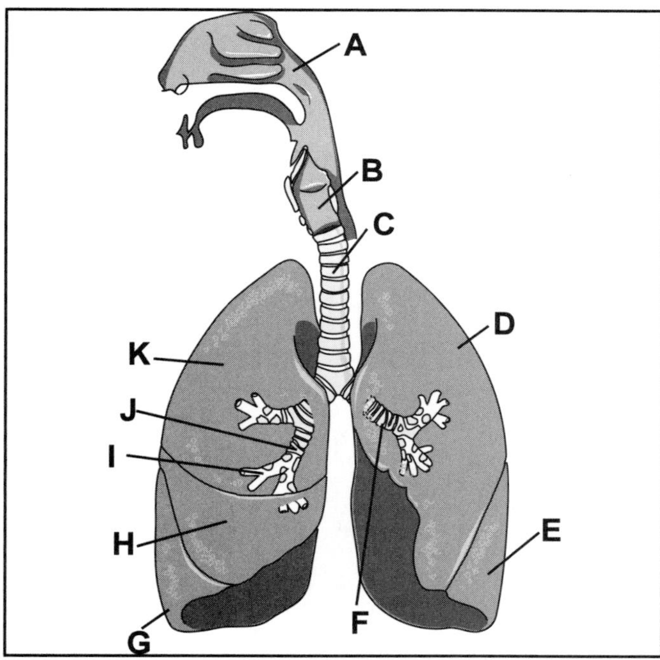

Fig. 16-3

9. What are the layers of serous membrane that protect the lungs?

10. Label the following diagram of alveoli.

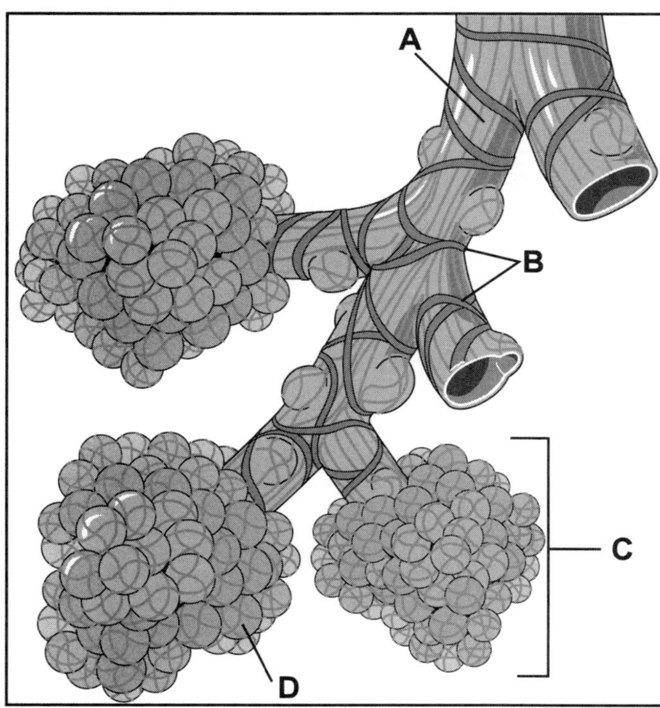

Fig. 16-4

11. The lungs are said to have a double blood supply. What does this mean and why is it necessary?

12. What is Boyle's law? How does it relate to pulmonary ventilation?

13. What produces pulmonary ventilation? What muscles are involved?

Questions 14–25 are matching. Match the following spirometry measure with its definition.

14.	tidal volume	a.	Volume of air that remains in the lungs and cannot be measure by spirometry
15.	anatomic dead space	b.	The volume of one normal breath
16.	minute ventilation rate	c.	The volume of air per minute that reaches the alveoli
17.	alveolar ventilation rate	d.	The volume of air that can be expelled by maximally effort in one second following a maximal inhalation
18.	inspiratory reserve volume	e.	Total inspiratory capacity of the lungs
19.	expiratory reserve volume	f.	Tidal volume multiplied by normal breathing rate
20.	forced expiratory reserve volume	g.	Sum of the inspiratory reserve volume, tidal volume, and expiratory reserve volume
21.	residual volume	h.	The volume of additional air that can be forcibly inhaled following a normal inhalation
22.	inspiratory capacity	i.	Sum of all lung volumes

23. functional residual capacity — j. Volume of air that remains in nose, pharynx, larynx, trachea, bronchi, bronchioles, and terminal bronchioles

24. vital capacity — k. Sum of residual volume and expiratory reserve volume

25. total lung capacity — l. Volume of air that can be forcibly exhaled following a normal exhalation

26. What is Dalton's law of partial pressures? How does it relate to gas exchange?

27. What is Henry's law? How does it relate to gas exchange?

28. What causes oxygen to enter the pulmonary systemic capillaries from the alveoli? What causes carbon dioxide to enter the alveoli from the pulmonary systemic capillaries?

29. What factors affect the rate of diffusion of gases in external respiration?

30. What causes oxygen to enter the tissues from pulmonary systemic capillaries? What causes carbon dioxide to the pulmonary systemic capillaries from the tissues?

31. How is oxygen transported in the blood?

32. What factors affect the affinity of hemoglobin for oxygen?

33. How is carbon dioxide transported in the blood?

34. What is the chloride shift?

35. What is the Haldane effect?

36. The respiratory center of the medulla oblongata and pons have three primary divisions. What are these areas and their basic functions?

37. What effect do cortical influences have on respiratory control?

38. What chemical regulation occurs for respiratory control?

Check Yourself

1. a) Frontal sinus

   b) Superior concha

   c) Middle concha

   d) Inferior concha

   e) Spenoidal sinus

   f) Pharyngeal tonsil

   g) Auditory tube

   h) Uvula

   i) Palatine tonsil

   j) Laryngopharynx

   k) Epiglottis

   l) Vocal fold

   m) Cricoid cartilage

   n) Trachea

   o) Thyroid cartilage

   p) Hyoid bone

   q) Tongue

   r) External napes

   s) Naso pharynx **(Anatomy of the respiratory system)**

2. The internal structures of the nose function to:

   a. warm, moisten, and filter incoming air.

   b. contain receptors of the olfactory system.

   c. produces a resonating chamber for vocalization. **(Physiology of the respiratory system)**

3. The mucus membrane contains capillaries and pseudostratified ciliated columnar epithelium, richly supplied with goblet cells. The mucus secreted along the surface of the epithelium moistens the air and traps foreign particles. The incoming air is also warmed by the proximity of the capillary networks. **(Physiology of the respiratory system)**

4. a) Epiglottis

   b) Hyoid bone

   c) Thyrohyoid membrane

   d) Thyroid cartilage

   e) Vestibular fold

   f) Vocal fold (Vocal chord)

   g) Cricoid cartilage

   h) Arytenoid cartilage

   i) Cornicullate cartilage **(Anatomy of the respiratory system)**

5. The epiglottis is a large flap of elastic cartilage covered by epithelium that serves as a lid over the glottis. The base of the epiglottis is attached to the thyroid cartilage, but the superior rim is unattached and free to move. During swallowing, as the larynx rises, the epiglottis is forced inferiorly to effectively seal the rima glottis preventing foods or liquids from entering the pharynx. **(Anatomy of the respiratory system)**

6. Vocalizations are produced when exhaled air is forced over the vocal folds (vocal cords) in the glottis. The vocal cords are bands of elastic ligaments stretched between the anterior and posterior plates of thyroid cartilage. Intrinsic muscles of the larynx attach to the cartilage and vocal folds such that contractions of the muscles produce tension on the vocal cords. Air rushing across the vocal cords during exhalations set up vibrations that travel as sound waves up the column of air. The greater the pressure of air, the more intense the sound. The greater the muscular contraction, the tighter the vocal cords become and the more rapidly they vibrate, producing higher frequency sound (pitch). **(Vocalization)**

7. The trachea is lined with 16–20 C-shaped rings of tracheal cartilage. These rings of rigid cartilage prevent the tracheal from collapse thus safeguarding the flow of air into the lungs. **(Anatomy of the respiratory system)**

8. a) Nasopharynx

   b) Trachea

   c) Tracheal cartilage

   d) Superior lobe of left lung

   e) Inferior lobe of left lung

   f) Primary bronchus

   g) Inferior lobe of right lung

   h) Middle lobe of right lung

   i) Tertiary bronchus

   j) Secondary bronchus

   k) Superior lobe of right lung **(Anatomy of the respiratory system)**

9. There are two layers of serous membranes surrounding the lungs, collectively called the pleural membrane. The most superficial layer lines the walls of the thoracic cavity and is called the parietal pleura. The deeper layer is the visceral pleura which lines the lung tissue. The space between the two is the pleural cavity, filled with a lubricating fluid. **(Anatomy of the respiratory system)**

10. a) Terminal bronchiole

    b) Pulmonary arteriole and venules

    c) Alveoli

    d) Alveolar cappillary **(Anatomy of the respiratory system)**

11. Deoxygenated blood is carried through the pulmonary trunk to eventually produce the alveolar capillaries for the exchange of gases. In addition, bronchial arteries branch from the aorta to supply the oxygenated blood to the tissues of the lungs. Most of this blood perfuses through the walls of the bronchi and bronchioles. **(Anatomy of the respiratory system)**

12. Boyle's law states that the pressure of a gas is inversely proportional to the volume of its container. Thus, if the volume of a container increases, the pressure of the gas contained in that pressure will decrease and vice versa. **(Pulmonary ventilation)**

13. Pulmonary ventilation (breathing) is accomplished by varying the volume of the lungs. The parietal pleura adheres to the thoracic cavity walls and the visceral pleura adheres to the lung tissue. The parietal and visceral pleurae adhere strongly to one another due to subatmospheric pressure between them, as well as the surface tension provided by the fluid filling the pleural cavity. For this reason, as the walls of the thoracic cavity move, lungs move in accordance. The base of the thoracic cavity is the dome-shaped diaphragm muscle. When it contracts, the diaphragm moves inferiorly as it flattens. At the same time, the external intercostal muscles contract pulling the ribs in a superior direction. The walls of the thoracic cavity are moved upward as the floor of the cavity moves downward resulting in an increase in the volume of the lungs. This increase lowers pressure inside the lungs (Boyle's law) to a level below atmospheric pressure causing air to rush into the lungs. Relaxation of these same muscles decreases the volume of the lungs, increasing pressure above that of the atmospheric pressure, and air moves out of the lungs. **(Pulmonary ventilation)**

14. b **(Lung volumes and capacities)**

15. j **(Lung volumes and capacities)**

16. f **(Lung volumes and capacities)**

17. c **(Lung volumes and capacities)**

18. h **(Lung volumes and capacities)**

19. l **(Lung volumes and capacities)**

20. d **(Lung volumes and capacities)**

21. a **(Lung volumes and capacities)**

22. e **(Lung volumes and capacities)**

23. k **(Lung volumes and capacities)**

24. g **(Lung volumes and capacities)**

25. i **(Lung volumes and capacities)**

26. Dalton's law of partial pressures states that each gas in a mixture of gases exerts its own pressure (p). The total pressure of a gas can be determined by adding the partial pressures of each gas in the mixture. This is a critical concept for gas exchange. Atmospheric air contains a variety of gases, primarily nitrogen, oxygen, carbon dioxide, water vapor, and some trace gases. In order for passive diffusion to power the gas exchange of oxygen from the atmosphere to the blood, the pO_2 must be greater in the atmosphere than in the blood. Similarly, for diffusion to power the exchange from the blood to the tissues, the pO_2 must be greater in the blood than in the tissue. The same holds true for the diffusion of carbon dioxide, but to eliminate carbon dioxide from the body, the pCO_2 needs to be greater in the tissues than in the blood, and again the pCO_2 of the blood must be greater than that of atmospheric for external respiration. **(Gas exchange)**

27. Henry's law states that the quantity of a gas that will dissolve in a fluid is proportional to the partial pressure of the gas above the fluid and the coefficient of solubility for that gas. The solubility coefficient for oxygen is relatively low (0.024) while the solubility coefficient for carbon dioxide is quite high (0.57). Henry's law indicates that the amount of gas diffusion for carbon dioxide and oxygen will not be solely due to the partial pressures of the gas, but will be effected by how soluble it is in the blood. This explains the fact that under normal conditions very little nitrogen exchange occurs from the atmosphere to our blood as we breathe. Although 79 percent of the partial pressure of air are due to nitrogen gas, its solubility is so low (0.012) that not much exchange can occur. **(Gas exchange)**

28. At sea level, the pCO_2 in the atmosphere is 0.3 mm Hg and the pO_2 is at about 160 mm Hg. Additional water vapor added to the air as it is moistened upon entering the body lowers the pO_2 to about 105 mm Hg. Since pCO_2 is so low, it is virtually unaffected by this process. However, due to the residual air in which gas exchange has already occurred that remains in the alveoli during expiration, the pCO_2 is greatly elevated at the alveoli (40 mm Hg) and the pO_2 is also slightly lowered (100 mm Hg). Blood returning from the tissues has lost oxygen to the cells and gained carbon dioxide from the cells. The pCO_2 in the blood entering the alveolar capillaries will be 45 mm Hg and the pO_2 will be 40 mm Hg. Due to the large numbers of capillaries in the pulmonary circuit, the blood flows relatively slowly, even during rigorous exercise, allowing the blood to come to equilibrium with the alveoli. Thus, blood flowing away from the alveoli in the pulmonary capillaries will have gained oxygen and lost carbon dioxide, with pCO_2 of 40 mm Hg and pO_2 of 100 mm Hg. **(Gas exchange)**

29. In external respiration, there are four principal factors that affect diffusion of gases:

    a. the difference in the partial pressures of gases,

    b. surface area of exchange,

    c. diffusion distance, and

    d. solubility and molecular weight of gases. **(Gas exchange)**

30. Internal respiration occurs passively as does external respiration. Blood entering the systemic capillaries has a pCO_2 of 40 mm Hg and a pO_2 of 100 mm Hg. Cellular respiration within the tissues will produce pCO_2 levels of 45 mm Hg and a pO_2 of 40 mm Hg in the interstitial fluid around these cells. Passive diffusion occurs providing a flow of oxygen from the blood to the interstitial tissues and on to the cells and a flow of carbon dioxide from the interstitial fluid into the blood. **(Gas exchange)**

31. Due to the low solubility coefficient of oxygen, only a small amount is carried and dissolved in the watery plasma, about 1.5 percent. The remainder is carried by hemoglobin in the erythrocytes. Every 100 ml of oxygenated blood carries about 20 ml of oxygen with 0.3 ml dissolved and 19.7 bound to hemoglobin. **(Oxygen transport)**

32. The affinity of oxygen to hemoglobin is directly affected by four main factors:

    a. acidity — as pH decreases, the protein structure of hemoglobin becomes slightly denatured producing a lower affinity for oxygen and thus increasing its unloading (Bohr effect),

    b. partial pressure of carbon dioxide — carbon dioxide can also bind to hemoglobin producing a similar affect to that of acidity,

    c. temperature — increases in heat decrease the affinity of hemoglobin and increases its unloading, and

d. 2,3 biphosphoglycerate (formerly called diphosphoglycerate) — decrease the affinity of hemoglobin and increases unloading.

Each of these effects is common in respiring tissue. Respiration generates heat and increases both the carbon dioxide and acidity in the interstitial fluid of surrounding these cells. Erythrocytes produce BPG as they break down glucose for energy. The lowered affinity of hemoglobin for oxygen in the area or respiring tissues enhances the delivery of oxygen. **(Oxygen transport)**

33. Carbon dioxide is carried and dissolved in the watery plasma (7 percent), as carbaminohemoglobin (23 percent), and as bicarbonate ions (70 percent). The majority of carbon dioxide reacts with water as it enters the blood cells forming carbonic acid, dissociating to form H^+ and HCO_3-. **(Carbon dioxide transport)**

34. The chloride shift occurs as HCO_3- accumulates in the erythrocytes. It diffuses down its concentration gradient out of the erythrocytes into the plasma. This increase in negativity in the plasma produces an electric gradient that moves Cl^- into the erythrocytes. **(Carbon dioxide transport)**

35. The Haldane effect is essentially the Bohr effect in reverse. In the alveolar capillaries, as oxygen binds to hemoglobin in the erythrocytes, the hemoglobin becomes more acidic and less capable of binding to carbon dioxide, releasing carbon dioxide for exchange with the alveoli. **(Carbon dioxide transport)**

36. The divisions of the respiratory center are:

    a. medullary rhythmicity area — controls the basic rhythm of respiration through activation of the phrenic nerve.

    b. pneumotaxic area — limits the duration of inspiration, facilitating expiration onset.

    c. apneustic area — it prolongs inspiration, inhibiting expiration.

    (Control of respiration)

37. Breathing is under partial voluntary control. Cortical influences can delay breathing or voluntarily alter breathing patterns, allowing coughing, sneezing, crying, etc. **(Control of respiration)**

38. The respiratory system is sensitive to concentrations of H^+, pO_2, and pCO_2 in the blood. Chemoreceptors are located centrally in the medulla oblongata and peripherally in the aortic arch and carotid bodies. Slight increases in H^+ and pCO_2 produce a condition called hypercapnia causing a strong response in both central and peripheral chemoreceptors and results in the elevation of the ventilation rate. Although less sensitive, peripheral receptors will respond to decreases in pO_2 , called hypoxia. If pO_2 falls to levels of approximately 50 mm Hg, the chemoreceptors will respond elevating ventilation rates. **(Control of respiration)**

Grade Yourself

Circle the numbers of the questions you missed, then fill in the total incorrect for each topic. If you answered more than three questions incorrectly, you need to focus on that topic. (If a topic has less than three questions and you had at least one wrong, we suggest you study that topic also. Read your textbook, a review book, or ask your teacher for help.)

Subject: The Respiratory System

Topic	Question Numbers	Number Incorrect
Anatomy of the respiratory system	1, 4, 5, 7, 8, 9, 10, 11	
Physiology of the respiratory system	2, 3	
Vocalization	6	
Pulmonary ventilation	12, 13	
Lung volumes and capacities	14, 15, 16, 17, 18, 19, 20, 21, 22, 23, 24, 25	
Gas exchange	26, 27, 28, 29, 30	
Oxygen transport	31, 32	
Carbon dioxide transport	33, 34, 35	
Control of respiration	36, 37, 38	

The Digestive System

17

Brief Yourself

Nutrients provide the energy necessary to drive the chemical reactions required in living organisms. In addition, nutrient sources supply the chemicals necessary for building and maintaining the body tissues and sustain the chemicals required for all reactions. Nutrients are obtained as foods and as consumed are not in a suitable state for utilization. The digestive system is designed to perform two major functions. Food particles are broken down into molecules of sufficient size to cross the plasma membrane in the process called digestion. The passage of these particles into the blood or lymphatic system is called absorption.

The digestive system organs are divided into two groups; the gastrointestinal tract and the accessory structures. The gastrointestinal tract (sometimes called the alimentary canal) is essentially a continuous tube from the mouth to the anus. The mouth, pharynx, esophagus, stomach, small intestine, and large intestine are all considered structures of the gastrointestinal tract. Dietary materials are digested inside the lumen of this tract. The accessory structures aid mechanical or chemical digestion and include the teeth, tongue, salivary glands, liver, gallbladder, and pancreas. The tongue and teeth function in ingestion and mechanical digestion, while the other accessory structures provide secretions necessary from chemical digestion.

Test Yourself

1. What are the six basic digestive system processes?

2. Label the following diagram of a sectional view of the gastrointestinal tract (GI tract).

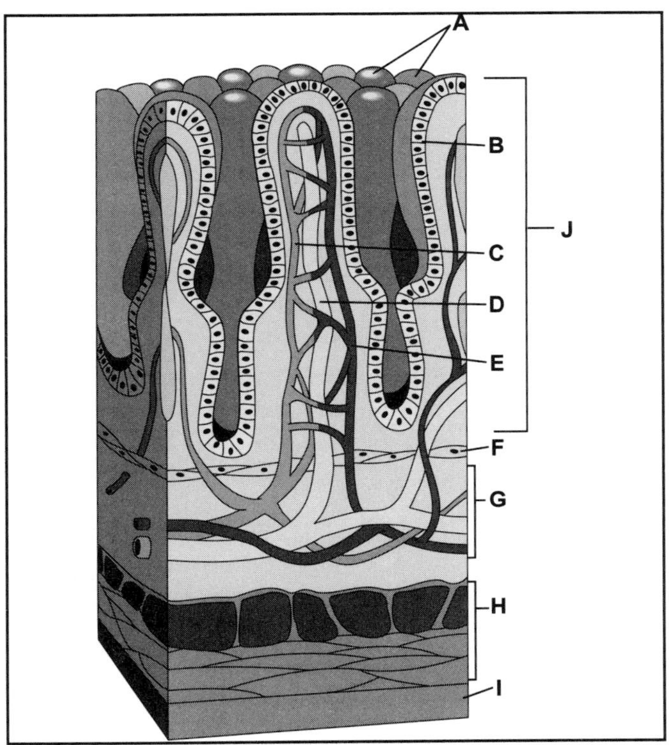

Fig. 17-1

3. What are the three layers that comprise the GI tract? What are the basic structure and function of each?

4. What is the peritoneum? What are its major structures?

5. What are deciduous teeth? What are permanent teeth and what are the four classes of permanent teeth?

6. What are salivary glands? What is the chemical nature of saliva?

7. What are the three principal parts of a tooth and what structures are found in each?

8. What digestion occurs in the mouth?

9. How does degluition occur?

10. What is the function of the esophagus? What is peristalsis?

11. What are the four principal areas of the stomach?

12. Label the following diagram of the stomach layers.

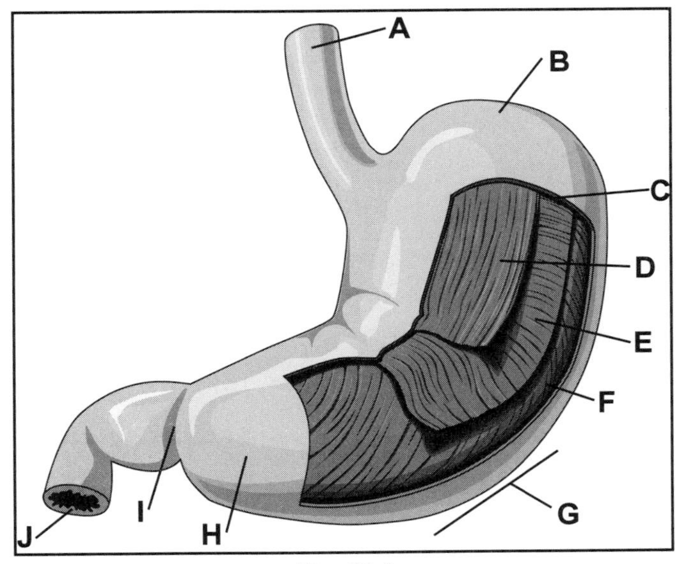

Fig. 17-2

13. What is the role of the stomach in mechanical digestion? What is its role in chemical digestion?

14. What is the cephalic phase of stomach activity?

15. What negative feedback cycle operates in the gastric phase of stomach activity?

16. The entry of chyme into the small intestine triggers the release of what three hormones? What is their effect on the stomach activity?

17. Label the following diagram of the liver.

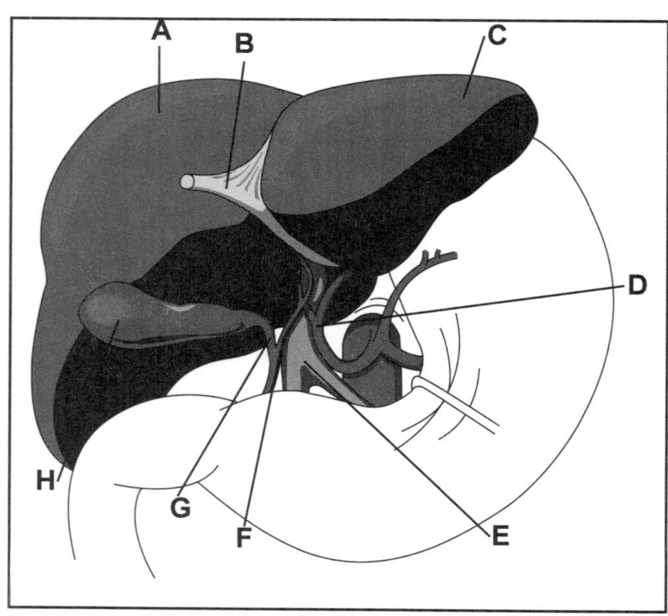

Fig. 17-3

18. What cells of the pancreas produce pancreatic juice? How is it delivered to the duodenum?

19. What enzymes are found in pancreatic juice?

20. Label the following diagram of a lobule of the liver.

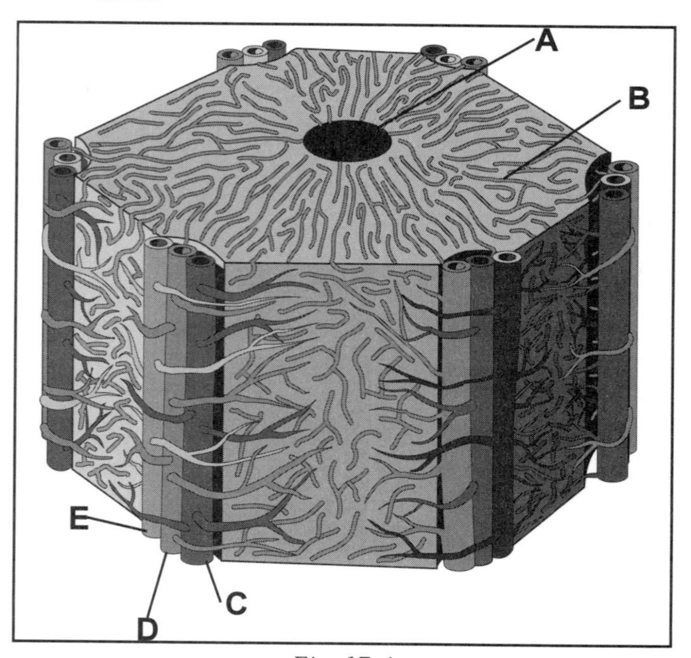

Fig. 17-4

21. What are the principal functions of the liver?

22. What is the digestive function of bile?

23. What is intestinal juice? What are brush border enzymes?

24. What two types of muscle contractions occur in the small intestine?

25. Describe carbohydrate digestion and absorption in the small intestine.

26. Describe protein digestion and absorption in the small intestine.

27. Describe lipid digestion and absorption in the small intestine.

28. In addition to carbohydrates, proteins, and lipids, what other substances are absorbed in the small intestine?

29. What mechanical digestion occurs in the large intestine? What chemical digestion occurs in the large intestine?

30. What regulates defecation?

Check Yourself

1. The basic digestive system processes include:

   a) ingestion

   b) secretion

   c) mixing and propulsion

   d) digestion (mechanical and chemical)

   e) absorption

   f) defacation **(Overview of digestive processes)**

2. a) Villi

   b) Absorptive (Epithelial) cell

   c) Venule

   d) Lacteal

   e) Arteriole

   f) Muscularis mucosae

   g) Submucosa

   h) Muscularis

   i) Serosa

   j) Mucosa **(Anatomy of the GI tract)**

3. The three layers of the GI tract are the mucosa, submucosa, and muscularis. The mucosa includes the epithelium, lamina propria, and muscularis mucosae. It is specialized for secretions and absorptions. The submucosa consists of areolar connective tissue that contains glands and lymphatic tissue. The muscularis is the muscular layer surrounding the GI tract. In the mouth, pharynx, and upper esophagus, it is primarily skeletal muscle, but throughout the remainder of the GI tract it is smooth muscle serving to mix food with digestive secretions and propel it along the tract. **(Anatomy of the GI tract)**

4. Peritoneum is the largest serous membrane of the body. The parietal peritoneum lines the wall of the abdominopelvic cavity and the visceral peritoneum covers all or part of the organs in the cavity. The space between the two layers is the peritoneal cavity and is filled with serous fluid. Unlike the serous membranes of the heart or lungs, the peritoneum contains large folds that weave between the visceral organs. These extensions bind the organs together and contain blood and lymph vessels and nerves to supply the abdominal organs. The major extensions include the mesentery, mesocolon, falciform ligaments, lesser omentum, and greater omentum. **(Anatomy of the GI tract)**

5. Deciduous teeth (milk teeth) are the primary dentation, appearing at approximately six months. They are replaced by the permanent teeth as they enlarge and develop. Typically, the replacement of deciduous teeth occurs between the ages of 6 and 12 years. The four classes of permanent teeth are incisors, canine, premolars, and molars. **(Anatomy of the mouth)**

6. Salivary glands are accessory structures of the digestive system that lie outside the mouth. They produce and secrete saliva, delivering it by ducts into the oral cavity. Saliva is 99 percent water and 1 percent solute. The solutes include ions, urea and uric acid, serum albumin and globulin, mucus, lysozyme, and salivary amylase. **(Salivary glands)**

7. The tooth is composed of the crown, neck, and body. The crown is visible above the gums and is composed of dentin and the outer layer of enamel. The neck is the juncture of the crown and the root. The roots are composed of dentin bounded by cementum, with a pulpy center. In the pulp cavity, blood vessels and nerves extend through the root canal to serve the tooth. The root is anchored by the periodontal ligament. **(Anatomy of the mouth)**

8. Mastication of food is the first phase of mechanical digestion. Chemical digestion also begins in the mouth by the action of salivary amylase that initiates the breakdown of starch. Small amounts of lingual lipase are also secreted in the saliva to initiate the digestion of dietary triglycerides into fatty acids and monoglycerides. **(Physiology of the mouth)**

9. Degluition (swallowing) is a muscular action that begins when the bolus of food is forced to the back of the oral cavity by the tongue. This voluntary stage of swallowing is followed by the involuntary pharyngeal stage of swallowing. As stretch receptors in the pharynx are stimulated, their activity results in smooth muscle contractions to close the respiratory passage, temporarily halt breathing, and elevate the larynx. Peristalsis in the muscularis of the pharynx, and then the esophagus, propels the bolus toward the stomach in the esophageal stage. **(Physiology of the mouth)**

10. The esophagus produces mucus and transports food to the stomach. Peristalsis is the involuntary contractions of the circular fibers of the muscularis of the GI tract that propel nutrients along the tract. **(Physiology of the esophagus)**

11. The principal areas of the stomach are the cardia, fundus, body, and pylorus. The cardia surrounds the superior opening to the stomach. The rounded superior extension of the stomach just superior to the cardia is the fundus. The body is the large, central portion of the stomach that extends from the cardia and fundus to the pylorus which connects to the duodenum. The pylorus contains the pyloric sphincter. **(Anatomy of the stomach)**

12. a) Esophagus

    b) Fundus

    c) Serosa

    d) Longitudinal muscularis

    e) Circular muscularis

    f) Oblique muscularis

    g) Greater curvature

h) Pylorus

i) Pyloric sphincter

j) Duodenum (**Anatomy of the stomach)**

13. Shortly after food enters the stomach, slow peristaltic movements called mixing waves mix the food with secretions of the gastric glands reducing the food to chyme. The secretions by parietal cells of H^+ and Cl^- produce a strong acidity to gastric fluid that begins to denature proteins in the food, as well as the protein salivary amylase and lingual lipase, halting their action. Chief cells secrete pepsinogen into the stomach which is converted to pepsin by the HCl. Pepsin is a strong proteolytic enzyme that breaks peptide bonds, converting long proteins into shorter fragments. Gastric lipase is also secreted, but has a limited effect due to the acidity in adult stomachs. **(Physiology of the stomach)**

14. The cephalic phase is triggered by sensory receptors that respond to the sight, smell, taste, and even the thought of food. The parasympathetic system responds by activating the secretions from all gastric glands, preparing the stomach to accept food. **(Physiology of the stomach)**

15. In the gastric phase, food entering the stomach disrupts stomach homeostasis, causing distention of the walls of the stomach and buffering stomach acid to increase pH. Chemoreceptors and stretch receptors lining the stomach are activated sending information to the submucosal plexus. Parasympathetic responses activate parietal cells to secrete more HCl and cause the muscularis in the walls of the stomach to contract. This increases the stomach acidity and lowers the distention of the stomach, returning to homeostasis. **(Physiology of the stomach)**

16. The entry of chyme into the duodenum triggers the release of gastric inhibitory peptide (GIP), secretin, and cholecystokinin (CCK). GIP inhibits the production of both gastric juices and gastric motility. Secretin also inhibits gastric secretions and CCK primarily inhibits stomach emptying. **(Physiology of the stomach)**

17. a) Right lobe

    b) Falciform ligament

    c) Left lobe

    d) Hepatic artery

    e) Hepatic portal vein

    f) Common hepatic duct

    g) Cystic duct

    h) Gall bladder **(Anatomy of the pancreas) (Anatomy of the liver)**

18. The pancreatic juice is produced by acinar cells in the exocrine function of the pancreas. The clusters of acinar cells (acini) lead to small ducts that converge to form the pancreatic duct that joins the common bile duct to form the hepatopancreatic ampulla that opens into the duodenum. **(Anatomy of the pancreas)**

19. Pancreatic juice includes pancreatic amylase, trypsin, chymotrypsin, carboxypeptidase, elastase, pancreatic lipase, ribonuclease, and deoxyribonuclease. Trypsin is secreted in an inactive form,

trypsinogen. Acinar cells also secrete a protein called trypsin inhibitor to prevent its activation prior to reaching the duodenum. **(Physiology of the pancreas)**

20. a) Central vein

    b) Hepatocytes

    c) Branch of hepatic artery

    d) Branch of hepatic portal vein

    e) Bile duct **(Anatomy of liver)**

21. The principal functions of the liver include:

    a) carbohydrate metabolism

    b) lipid metabolism

    c) protein metabolism

    d) removal of drugs and hormones

    e) excretion of bilirubin

    f) synthesis of bile salts

    g) storage

    h) phagocytosis

    i) activation of vitamin D **(Physiology of the liver)**

22. Bile salts serve a role in the emulsification of lipids. It serves to break large lipid globules into small droplets, suspending them in the intestinal fluid and thus providing a large surface area for pancreatic lipase activity. **(Physiology of the liver)**

23. Intestine juice is the combination of chyme delivered from the stomach mixed with intestinal secretions, pancreatic juice, and bile. The absorptive epithelial cells lining the small intestine synthesized a number of digestive enzymes that are displayed on the plasma membranes of the apical faces of their microvilli. These enzymes include dextrinase, maltase, sucrase, lactase, aminopeptidase, dipeptidase, nucleosidase, and phosphatase. **(Physiology of the small intestine)**

24. Two types of muscular contractions that occur in the small intestine are segmentation and peristalsis. Segmentation is localized strong contractions that serve to mix chyme and digestive juices, bringing nutrient particles in contact with the mucosa for absorption. Peristalsis are weak contractions that propel the chyme through the small intestine and ultimately into the large intestine. **(Physiology of the small intestine)**

25. Although carbohydrate digestion begins in the mouth with salivary amylase, the major digestion occurs in the small intestine due to the actions of pancreatic amylase. The brush border enzymes dextrinase, maltase, sucrase, and lactase complete the digestion of carbohydrates into monosaccharides for absorption. The movement of monosaccharides occurs across the plasma membrane of the apical face of

epithelial cells lining the small intestine lumen. This movement occurs by facilitated diffusion or active transport (coupled with Na^+ transport). **(Physiology of the small intestine)**

26. Protein digestion begins with the activity of pepsin and acid in the stomach and continues in the small intestine by the actions of pancreatic enzymes trypsin, chymotrypsin, and elastase. The resulting small peptides react with pancreatic carboxypeptidase that cleaves the terminal amino acid from the peptide chain. Brush border aminopeptidase and dipeptidase also serves to cleave single amino acids from peptide fragments. Amino acids are absorbed across the plasma membrane of the apical face of epithelial cells lining the lumen of the small intestine. Most amino acids are absorbed by active transport. **(Physiology of the small intestine)**

27. A very small amount of lipid digestion occurs in the mouth and stomach due to the activity of lingual and gastric lipases, but for adults the primary lipid digestion occurs in the small intestine. Bile salts from the liver emulsify the globules of triglycerides into small droplets in the small intestine increasing the surface area for the action of pancreatic lipase. Lipase breaks triglycerides into monoglycerides and fatty acids which are absorbed by simple diffusion, moving across the apical face of epithelial cells lining the lumen of the small intestine. **(Physiology of the small intestine)**

28. In addition to carbohydrates, lipids, and proteins, the small intestine absorbs water, ions (electrolytes), and vitamins. Roughly 89 percent of the water entering the small intestine is recovered due to osmosis as the water follows the electrolyte concentrations. Electrolytes, particularly Na^+ and Ca^{++}, are actively transported allowing chloride, iodide, and nitrates to follow passively. Fat-soluble vitamins are carried along with dietary lipids by passive diffusion. Water-soluble vitamins are also absorbed passively, except for vitamin B_{12}, which combines to intrinsic factor and is absorbed by receptor-mediated endocytosis. **(Physiology of the small intestine)**

29. The characteristic contractions of the large intestine include haustral churning, peristalsis, and mass peristalsis. Haustral churning occurs with the filling and distention of an individual haustrum. When sufficiently stretched, the muscularis of the haustrum contracts, forcing the contents into the next haustrum. A slow peristalsis occurs in the proximal portion of the large intestine, with stronger, mass peristaltic contractions occurring near the middle of the transverse colon that drives the colonic contents into the rectum. Although mucus is secreted by the glands of the large intestine, no enzymes are secreted. The chyme is processed by bacteria living in the large intestine, fermenting remaining carbohydrates and converting remaining proteins to amino acids for absorption. **(Physiology of the large intestine)**

30. Defecation is regulated by the defecation reflex. Distention of the rectal wall stimulates stretch receptors. Their activation results in increased contractions of the longitudinal rectal muscles, shortening the rectum and increasing the pressure. This pressure, coupled with voluntary contractions of the diaphragm and abdominal muscle expels feces through the anus. The external sphincter is voluntarily controlled. If relaxed, defecation occurs, but if it is constricted, defecation can be postponed. **(Physiology of the small intestine)**

Grade Yourself

Circle the numbers of the questions you missed, then fill in the total incorrect for each topic. If you answered more than three questions incorrectly, you need to focus on that topic. (If a topic has less than three questions and you had at least one wrong, we suggest you study that topic also. Read your textbook, a review book, or ask your teacher for help.

Subject: The Digestive System

Topic	Question Numbers	Number Incorrect
Overview of digestive processes	1	
Anatomy of the GI tract	2, 3, 4	
Anatomy of the mouth	5, 7	
Salivary glands	6	
Physiology of the mouth	8, 9	
Physiology of the esophagus	10	
Anatomy of the stomach	11, 12	
Physiology of the stomach	13, 14, 15, 16	
Anatomy of the pancreas, anatomy of the liver	17	
Anatomy of the pancreas	18	
Physiology of the pancreas	19	
Anatomy of liver	20	
Physiology of the liver	21, 22	
Physiology of the small intestine	23, 24, 25, 26, 27, 28, 30	
Physiology of the large intestine	29	

Metabolism and Nutrition

18

Brief Yourself

Metabolism refers to all of the chemical reactions that occur within a living organism. These reactions may be classified as anabolic, in which more complex molecules are synthesized from simpler precursors and catabolic, in which complex molecules are decomposed into simpler molecules. Overall, anabolic reactions are endergonic, consuming more energy than they produce. Catabolic reactions are exergonic, releasing more energy than they consume. Although there are exceptions, anabolic reactions are typically reduction reactions and catabolic reactions are typically oxidation reactions. Metabolism acts to maintain the balance between energy-producing catabolic and energy-requiring anabolic reactions so that the chemical processes necessary to life may be controlled.

Many of the molecules required to maintain or build our tissues can be synthesized within our bodies, but others cannot. These molecules, such as essential amino acids, essential lipids, vitamins, and minerals must be ingested and absorbed by the gastrointestinal tract. These essential molecules are known as nutrients. Most nutrients provide a usable chemical energy or building blocks for chemical synthesis.

Test Yourself

1. Why are anabolic and catabolic reactions said to be coupled by adenosine triphosphate (ATP)?
2. What is basal metabolic rate? How is it typically measured?
3. What is a calorie? How is caloric value of food determined?
4. What physical factors influence the metabolic rate and heat production in an individual?
5. If body temperature starts to decrease, what effectors cause its increase?
6. How is heat lost from the body?
7. What are oxidation-reduction (redox) reactions? What coenzymes are typically used in animal cells to carry hydrogen atoms?
8. What is phosphorylation? What three methods of phosphorylation are utilized to generate ATP?
9. What is glycolysis? What are its principal events and products?
10. How is acetyl coenzyme A formed? Where does this formation occur in cellular respiration?
11. What are the principal events and products of the Kreb's cycle?

12. What happens in the electron transport chain? What is chemiosmosis?

13. What are the advantages gained by performing aerobic respiration in mitochondria?

14. Define glycogenesis and glycogenolysis. When is each most likely to occur?

15. What is gluconeogenesis and why is it important?

16. How are triglycerides absorbed? How are they stored?

17. What are the principal events and products of the catabolism of glycerol and fatty acids?

18. What is ketosis and what causes it?

19. Define lipogenesis and lipolysis. When is each most likely to occur?

20. What is deamination? What happens to the ammonia produced in this process?

21. In order to be catabolized to produce energy, amino acids must be converted to substrates of the Kreb's cycle. Identify the amino acid which can be converted into the following substrates:

    a. pyruvic acid
    b. acetyl coenzyme A
    c. oxaloacetic acid
    d. fumaric acid
    e. succinyl coenzyme A
    f. alpha-ketoglutaric acid

22. What is the absorptive state? What are its principal events?

23. What is the postabsorptive state? What are its principal events?

24. What is a mineral? What are the general functions of minerals in the body?

Questions 25–40 are matching. Match the following mineral with its function:

25. Calcium	a. Plays a role in acid-base balance, water balance, formation of HCl in stomach
26. Phosphorus	b. Part of B_{12}, required for erythropoiesis
27. Iron	c. Formation of bones, teeth, blood clotting, muscle and nerve activity, endocytosis and exocytosis, cellular motility, chromosome movement, glycogen metabolism, and synthesis and release of neurotransmitters
28. Iodine	d. Antioxidant, prevents chromosome breakage
29. Copper	e. Functions in nerve and muscle action potential conduction
30. Sodium	f. Activate enzymes, needed for hemoglobin synthesis, urea formation, growth, reproduction, lactation, bone formation, and inhibition of cell damage
31. Potassium	g. Formation of bone and teeth, formation of phosphates, plays a role in muscle and nerve activity, component of many enzymes, component of DNA and RNA, essential for energy transfer
32. Chlorine	h. As a component of hemoglobin binds oxygen, component of cytochromes in electron transport chain

33. Magnesium
34. Sulfur
35. Zinc
36. Flourine
37. Manganese
38. Cobalt
39. Chromium
40. Selenium

i. Component of hormones and vitamins, needed for ATP production in aerobic respiration

j. Required with iron for hemoglobin synthesis, component of coenzymes in electron transport chain, necessary for melanin formation

k. Affects distribution of water through osmosis, component of bicarbonate buffers, plays a role in muscle and nerve action potential conduction

l. Required for synthesis of thyroid hormones

m. Inhibits tooth decay

n. Required for muscle and nerve function, participates in bone formation, component of many enzymes

o. Component of carbonic anhydrase, important to carbon dioxide metabolism, necessary for growth and wound healing, required for sperm production in males, required for taste sensation, involved in protein digestion

p. Needed for proper use of dietary sugars, effects insulin production, increase HDL and decrease LDL

41. What is a vitamin? How do fat-soluble and water-soluble vitamins differ?

Questions 42–54 are matching. Match the following vitamin with its function:

42. Vitamin A
43. Vitamin D
44. Vitamin E
45. Vitamin K
46. Vitamin B_1
47. Vitamin B_2
48. Niacin
49. Vitamin B_6

a. Promotes many metabolic reactions, particularly protein metabolism (especially collagen), promotes antibody actions, promotes healing, functions as antioxidant

b. Coenzyme essential to blood clotting

c. Component of coenzyme A, plays a role in synthesis of cholesterol and steroid hormones

d. Essential component of NAD, inhibits production of cholesterol

e. Maintains health of epithelial cells, acts as antioxidant, plays a role in formation of photopigments, aids in bone and teeth growth by regulating osteoblast and osteoclast activity

f. Coenzyme necessary for erythrocyte formation, formation of methionine, formation of choline, and entry of certain amino acids into Kreb's cycle

g. Coenzyme for conversion of pyruvic acid to oxaloacetic acid

h. Essential for absorption and utilization of calcium and phosphorus from GI tract, works with parathormone to maintain calcium levels

50.	Vitamin B_{12}	i.	Component of enzyme system for synthesizing purines and pyramidines for DNA and RNA, essential for hematopoiesis
51	Panto-thenic acid	j.	Coenzyme for many enzymes that break carbon-carbon bonds, essential for synthesis of acetylcholine
52.	Folic acid	k.	Necessary for amino acid metabolism, assists in circulation of antibodies
53.	Biotin	l.	Inhibits catabolism of fatty acids to aid formation of membranes, involved in DNA and RNA formation, involved in erythropoiesis, act as antioxidant and may promote wound healing
54.	Vitamin C	m.	Component of coenzymes FAD and FMN, found in large amounts in cells of eye, integument, and mucosa

Check Yourself

1. Anabolic reactions are endergonic, requiring energy to complete. Catabolic reactions are exergonic, releasing energy as they occur. In order to produce our anabolic reactions, our cells must have a usable chemical energy source, ATP. The energy released by catabolism of complex nutrients such as carbohydrates, lipids, or proteins is transferred to ATP in its production and then utilized to power anabolic reations. **(Metabolism)**

2. Basal metabolic rate is the measurement of the rate at which the resting body breaks down nutrients to release energy. It is most typically measured indirectly by spirometry, which determines the amount of oxygen uptake by the body. Nutrients require known amounts of oxygen to combine in order to release their energy. By measuring the amount of oxygen consumed, an accurate estimate of the amount of released energy may be obtained. **(Metabolism)**

3. A calorie is the amount of heat (kinetic) energy that is required to raise the temperature of 1.0 gram of water from 14°C to 15°C. Since this is a small amount relative to the amount of energy stored in foods, the typical measure utilized to express food energy is a kilocalorie (or Calorie spelled with a capital C) equal to 1000 calories. The caloric value of food is measured by burning the food in a calorimeter and measuring the temperature increase in water surrounding it. **(Metabolism)**

4. Metabolic rate and heat production are affected by:

   a. exercise — exercise dramatically increases metabolic rate.

   b. hormones — thyroid hormones are the main controllers of metabolic rate.

   c. nervous system — during stress, the sympathetic nervous system release norepinephrine and stimulates the endocrine system to release both epinephrine and norepinephrine, all of which increase metabolic rate.

   d. body temperature — the higher the temperature, the higher the metabolic rate.

   e. ingestion of food — the specific dynamic effect of foods increases metabolic rate.

   f. age — children's metabolic rate is higher than that of adults, primarily due to the rates of reactions required for growth.

   g. gender — males metabolic rate is typically greater than females (except during pregnancy and lactation).

   h. climate — cooler climates require a greater metabolic rate.

   i. arousal — during sleep, metabolism is slower.

   j. malnutrition — malnutrition lowers metabolic rate. **(Body temperature)**

5. Body temperature decreases will be sensed by thermoreceptors in the skin and hypothalamus, activating the hypothalamic heat-promoting center. The hypothalamus will release thyrotropin-releasing hormones producing the following changes:

a. sympathetic stimulation to produce vasoconstriction of blood vessels in the skin, decreasing the loss of heat from the blood at the surface of the body.

b. sympathetic stimulation and release of epinephrine and norepinephrine to increase metabolism.

c. skeletal muscle activation to produce shivering.

d. thyroid hormones are released to increase metabolism. **(Body temperature)**

6. Heat is lost from the body by radiation, evaporation (from perspiration), conduction, and convection. **(Body temperature)**

7. Oxidation is the removal of electrons from a molecule resulting in a decrease of energy content in that molecule. The addition of electrons to a molecule resulting in an increase of energy content of a molecule is reduction. Since electrons are transferred, whenever one molecule is oxidized, another is simultaneously reduced and the reactions are said to be coupled (redox). Most biological oxidations involve the removal of hydrogen atoms. The two coenzymes that typically carry hydrogen during its transfer are nicotinamide adenine dinucleotide (NAD) and flavin adenine dinucleotide (FAD). **(Energy production)**

8. Phosphorylation is a chemical reaction that results in the addition of a phosphate (PO_4-3) to a molecule. In cellular respiration, the production of adenosine triphosphate from the addition of a PO_4-3 to adenosine diphosphate is a key reaction. It may be accomplished by substrate phosphorylation or oxidative phosphorylation. **(Energy production)**

9. Glycolysis is a series of ten chemical reactions that occur anarobically resulting in the decomposition of a 6-carbon glucose molecule into 2 3-carbon pyruvic acid molecules. The principal events and products of glycolysis are as follows:

   a. the first three reactions serve to add two PO_4–3 to the substrate sugar converting glucose to fructose 1,6 diphosphate, requiring the conversion of two molecules of ATP to ADP.

   b. in the fourth reaction, fructose 1,6 diphophate is cleaved producing two 3-carbon compounds, one glyceraldehyde 3-phosphate (G 3–P) and one dihydroxyacetone phosphate which converts to G 3-P in the fifth reaction.

   c. in the sixth reaction, oxidation occurs as 2 NAD^+ accepts two pairs of electrons and two hydrogen ions from the two molecules of G 3–P forming two reduced $NADH+H^+$ and converting the two G 3-P molecules into two molecules of 1,3 diphosphoglyceric acid (each gaining an inorganic PO_4–3).

   d. in the remaining 4 four reactions, substrate phoshorylation is responsible for the conversion of four ADP molecules to four ATP molecules and the coversion of the two 1,3 diphosphoglyceride molecules into two molecules of pyruvic acid The final products of glycolysis are 2 molecules of pyruvic acid, 4 molecules of ATP (with a net gain of 2), and 2 molecules of $NADH+H^+$. **(Energy production)**

10. If oxygen is plentiful in a cell undergoing glycolysis (aerobic), the two pyruvic acid molecules will be transported by special carriers in the plasma membrane of the mitochondria into the mitochondrial matrix. Upon entering the mitochondrial matrix, they are decarboxylated (a molecule of CO_2 is released from each). During the same reaction, each will lose electrons and hydrogen ions resulting in the reduction of two molecules of NAD^+ to two molecules of $NADH+H^+$. The resulting 2-carbon molecules are acetyl groups which then combine with two coenzyme A molecules resulting in the formation of two molecules of acetyl coenzye A. **(Energy production)**

11. The Kreb's cycle is a series of nine reactions that occur in the matrix of the mitochondria that begin with the combination of acetyl coenzyme A with oxaloacetic acid, resulting in the release of coenzyme A and the formtion of citric acid. The final reaction of Kreb's cycle will result in the formation of oxaloacetic acid and for this reason it is designated a "cyclic" reaction. The principal events and products of the Kreb's cycle are as follows:

    a. the second and third reactions transform citric acid to isocitric acid.

    b. the fourth reaction is a decarboxylation and transformation to alpha-ketoglutaric acid providing for the reduction of NAD to $NADH+H^+$.

    c. the fifth reaction combines coenzyme A with alpha-ketoglutaric acid to produce succinyl coenzyme A, again resulting in a decarboxylation and reduction of NAD to $NADH+H^+$.

    d. the sixth reaction reaction releases coenzyme A, converting succinyl coenzyme A to succinic acid, powering the phosphorylation of guanosine diphosphate to guanosine triphosphate which in turn provides the substrate phosphorylation of ADP to ATP.

    e. the seventh reaction converts succinic acid to fumaric acid allowing the reduction of FAD to $FADH_2$.

    f. the eight reaction converts fumaric acid to malic acid.

    g. the final reaction converts fumaric acid to oxaloacetic acid allowing the reduction of NAD to $NADH+H^+$.

    Two molecules of acetyl coenzyme A are produced from the two molecules of pyruvic acid produced in glycolysis. Thus, for each glycolysis the Kreb's cycle occurs twice. The total products of two Kreb's cycles are 2 molecules of ATP by substrate phosphorylation, 6 molecules of $NADH+H^+$, and two molecules of $FADH_2$. **(Energy production)**

12. Electron transport involves a chain of electron carrier molecules on the inner membrane of the mitochondria that produce oxidation-reduction reactions. As electrons pass along the chain, energy is released that will serve to power the phosphorylation of ADP to ATP. The final reservoir for the electrons is oxygen and for this reason this type of phosphorylation is called oxidative phosphorylation. Because this type of ATP production uses chemical reactions to pump electrons, it is called chemiosmosis. Energy from the $NADH+H^+$ and $FADH_2$ is used to pump H^+ across the inner membrane into the outer compartment of the mitochondria, producing a high concentration of H^+ in this compartment. Due to the resulting electrical and chemical gradients, the H^+ will diffuse back across the inner membrane through an ATP-synthase channel that acts to catalyze the phosphorylation of ADP to ATP. The H^+ will join with oxygen to form water. **(Energy production)**

13. The mitochondria contains an inner and outer membrane producing separate compartments within the mitochondria. This allows the development of concentration and electrical differences across the inner membrane between the two compartments, producing the capability for the ionic current that finally powers oxidative phosphorylation. **(Energy production)**

14. Glycogenesis is the process of combining glucose molecules into long chains producing glycogen. Glycogenolysis is the decomposition of glycogen into its component glucose molecules. Glycogenesis typically occurs when there is abundant glucose so that more is available than is required for ATP production. It is stimulated by insulin released by the pancreas. Glycogenolysis occurs when glucose levels are low, typically between meals and is stimulated by glucagon released by the pancreas and epinephrine released by the adrenal medulla. **(Glucose metabolism)**

15. The production of glucose from noncarbohydrate sourses is called gluconeogenesis. The process occurs in the liver and can utilize lactic acid, certain amino acids, and the glycerol component of triglycerides as substrates. Gluconeogenesis protects the body, the nervous system in particular, from the damaging effects of low blood sugar by ensuring ATP synthesis can continue even when dietary and stored glucose has been depleted. **(Glucose metabolism)**

16. Triglycerides are digested into fatty acids and monoglycerides. The short-chain fatty acids diffuse into the epithelial cells of the GI tract and then into the capillaries. The long-chain fatty acids and monoglycerides are transported in micelles into the epithelial cells of the GI tract and then combined to form triglycerides to enter the lymphatic capillaries in chylomicrons, entering the blood stream through the thoracic duct. Triglycerides are primarily stored in the adipose tissue. **(Lipid metabolism)**

17. The first step in the catabolism of triglycerides is its decomposition into glycerol and three fatty acid chains. Glycerol can be readily converted into glyceraldehyde 3-phosphate and can enter the glycolysis pathway at that point. Fatty acids enter a series of reactions called beta oxidation which fragments the fatty acids into pairs of carbon atoms that attach to coenzyme A to form acetyl coenzyme A. The acetyl coenzyme A can enter Kreb's cyle at that point. **(Energy production)**

18. Ketosis is the accumulation of ketone bodies in the blood at abnormally high levels. As a normal part of fatty acid metabolism, hepatocytes condense two acetyl coenzyme A molecules into acetoacetic acid which often coverts to beta-hydroxybutyric acid and acetone. These substances are know as ketone bodies. During excessive beta oxidation, the ketone content in the blood will rise. **(Lipid metabolism)**

19. Lipogenesis is the process of synthesizing lipids from glucose or amino acids. Stimulated by insulin, this process typically takes place when there are more calories consumed than needed to for ATP production. Lipolysis is the process by which stored triglycerides are released and broken into glycerols and fatty acid chains. Lipolysis is enhanced by a variety of hormones including epinephrine, norepinephrine, cortisol, thyroid hormones, and human growth hormone. **(Lipid metabolism)**

20. As a part of protein catabolism, deamination is the removal of the amine group (NH_2) from the amino acid. The resulting ammonia (NH_3) is converted to urea which is then excreted in the urine. **(Protein metabolism)**

21. a. alanine, cysteine, glycine, serine, threonine

    b. isoleucine, leucine, tryptophan

    c. aspartic acid, aspargine

    d. tyrosine, phenylalanine

    e. isoleucine, methionine, valine

    f. arginine, histidine, glutamine, proline **(Protein metabolism)**

22. The absorptive state is the period following ingestion of nutrients as they enter your bloodstream. The principal events of the absorptive state are as follows:

    a. most cells are producing ATP by oxidizing glucose.

    b. glucose entering the hepatocytes is converted to triglycerides or glycogen.

    c. adipose cells take up glucose not converted in the liver and convert to triglycerides for storage.

d. skeletal muscle cells store glucose as glycogen.

e. most dietary lipids are stored in the liver or adipose tissue.

f. most dietary amino acids are taken up by the cells of the body for protein synthesis.

g. many amino acids entering the hepatocytes are deaminated to keto acids.

h. remaining amino acids entering the hepatocytes are used for protein synthesis. **(Absorptive state)**

23. The postabsorptive state occurs during the interval between ingestion of nutrients. The requirements of the body must be met by existing stores. The principal events of the postabsorptive state are as follows:

    a. glycogen in the liver is converted to glucose.

    b. lipolysis occurs in the adipose tissue releasing fatty acids and glycerol.

    c. gluconeogenesis occurs in the liver.

    d. oxidation of fatty acids, lactic acids, amino acids, and ketone bodies occurs.

    e. muscle glycogen is converted to glucose. **(Postabsorptive state)**

24. A mineral is an inorganic substance. It may appear as ions in solution or in combination with other organic compounds. Minerals serve as components in enzyme systems, produce electrolytes in the fluids of the body, form part of the structure of bone, and catalyze certain reactions. **(Minerals)**

25. c **(Minerals)**

26. g **(Minerals)**

27. h **(Minerals)**

28. l **(Minerals)**

29. j **(Minerals)**

30. k **(Minerals)**

31. e **(Minerals)**

32. a **(Minerals)**

33. n **(Minerals)**

34. i **(Minerals)**

35. o **(Minerals)**

36. m **(Minerals)**

37. f **(Minerals)**

38. b **(Minerals)**

39. p **(Minerals)**

40. d **(Minerals)**

41. Organic molecules needed in minute amounts in order to maintain normal metabolism in the body are called vitamins. Vitamins are classified according to their solubility. Fat-soluble vitamins are emulsified into micelles and absorbed along with the lipids in the diet. These vitamins are easily stored in the lipid component of cells, particularly hepatocytes. The vitamins A, D, E, and K are fat-soluble. Water-soluble vitamins are absorbed with the water in the GI tract and are not stored. The B vitamins and vitamin C are water soluble. **(Vitamins)**

42. e **(Vitamins)**

43. h **(Vitamins)**

44. l **(Vitamins)**

45. b **(Vitamins)**

46. j **(Vitamins)**

47. m **(Vitamins)**

48. d **(Vitamins)**

49. k **(Vitamins)**

50. f **(Vitamins)**

51. c **(Vitamins)**

52. i **(Vitamins)**

53. g **(Vitamins)**

54. a **(Vitamins)**

Grade Yourself

Circle the numbers of the questions you missed, then fill in the total incorrect for each topic. If you answered more than three questions incorrectly, you need to focus on that topic. (If a topic has less than three questions and you had at least one wrong, we suggest you study that topic also. Read your textbook, a review book, or ask your teacher for help.)

Subject: Metabolism and Nutrition

Topic	Question Numbers	Number Incorrect
Metabolism	1, 2, 3	
Body temperature	4, 5, 6	
Energy production	7, 8, 9, 10, 11, 12, 13, 17	
Glucose metabolism	14, 15	
Lipid metabolism	16, 18, 19	
Protein Metabolism	20, 21	
Absorptive state	22	
Postabsorptive state	23	
Minerals	24, 25, 26, 27, 28, 29, 30, 31, 32, 33, 34, 35, 36, 37, 38, 39, 40	
Vitamins	41, 42, 43, 44, 45, 46, 47, 48, 49, 50, 51, 52, 53, 54	

The Urinary System

19

Brief Yourself

The primary function of the urinary system is the maintenance of the composition, volume, and pressure of the blood. The urinary system consists of two kidneys, two ureters, one urinary bladder, and one urethra. The kidneys filter the blood, eliminating wastes and restoring water and necessary solutes to the bloodstream. The wastes are carried away in the fluid urine, excreted by each kidney through its ureter to be stored in the urinary bladder until it is expelled from the body through the urethra.

Wastes are substances in the body with no function or essential substances that may build up to excess concentrations. The kidney is responsible for the elimination of excess water and ions, as well as metabolic by-products such as ammonia, urea, bilirubin, and uric acid. In addition, some bacterial toxins, CO_2, and heat are eliminated by the urinary system. The kidneys also play essential roles in the regulation of blood volume and composition, regulation of blood pH, regulation of blood pressure, and contribution to metabolism.

Test Yourself

1. What are the four major organs of the urinary system?

2. Label the following diagram of the frontal sections of the kidney.

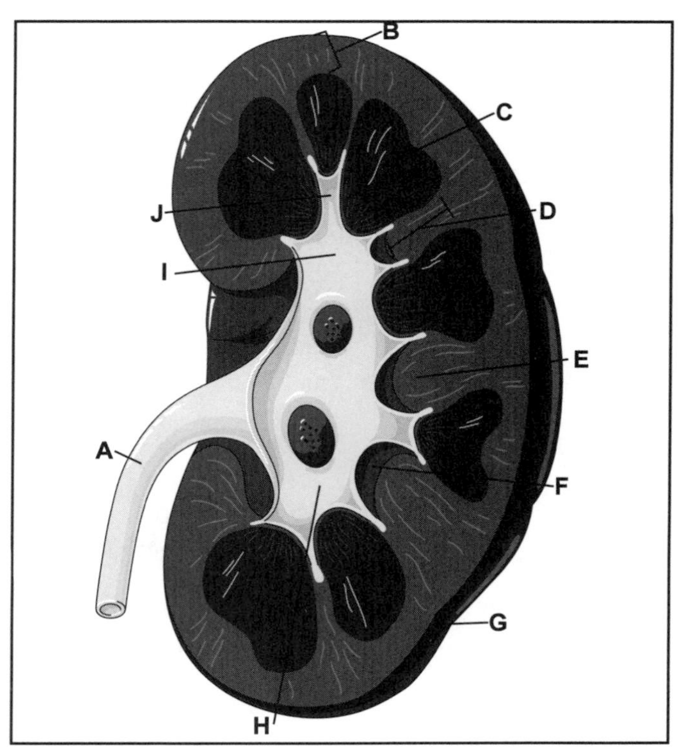

Fig. 19-1

3. Describe the location of the kidneys in the body. What layers of connective tissue protect and anchor the kidneys?

4. Beginning with the renal artery and ending with the renal vein, describe the circulatory system structures for the path of blood through the kidneys.

5. What is a podocyte? What is its function in the urinary system?

6. Label the following diagram of a nephron.

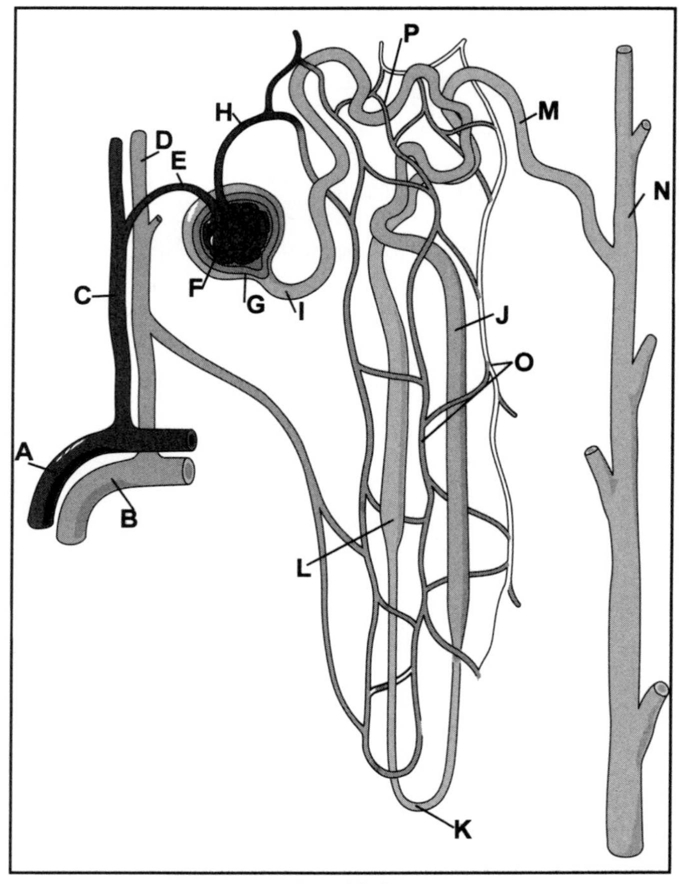

Fig. 19-2

7. Identify and describe the three basic functions of a nephron.

8. What is glomerular filtration? What features of the renal corpuscle enhance filtration?

9. What is net filtration pressure? How is it calculated?

10. What is renal autoregulation of glomerular filtration rate (GFR)? How does it occur?

11. What two hormones are most effective in controlling blood pressure? What are the effects of each?

12. What neural regulation of GFR occurs?

13. What is tubular reabsorption? Where is it most effective?

14. How is Na^+ reabsorption accomplished in the proximal convoluted tubule (PCT)? What is its effect on other ions and on water?

15. How is the reabsorption of nutrients accomplished in the PCT?

16. What reabsorption occurs in the loop of Henle?

17. What reabsorption occurs in the distal convoluted tubule (DCT) and collecting ducts? What hormones affect the activity of the DCT?

18. What is tubular secretion? What are the two principal effects of tubular secretion?

19. Identify the three principal substances secreted by a nephron. Describe the method and location of the secretions of each.

20. What are the three steps to the production of a more concentrated urine?

21. What is the countercurrent mechanism? How is it accomplished?

22. What is the function of the ureters? How is it accomplished?

23. Why is an anatomical valve not required at the entry of each ureter into the urinary bladder?

24. What is micturition? How is it controlled?

25. In males, what reproductive secretions are discharged through the urethra?

26. What organic solutes are typically found in urine? What inorganic solutes are typically found in urine?

Questions 27–33 are matching. Match the following condition with its cause.

27. Albuminuria	a. May be due to diabetes mellitus, anorexia, or starvation
28. Glucosuria	b. May be due to hemolytic or pernicious anemia, infectious hepatitis, biliary obstruction, jaundice, cirrhosis, heart failure or infectious mononucleosis
29. Hematuria	c. May be due to injury or disease, increased blood pressure, irritation of kidney cells, or heavy metals
30. Pyuria	d. May be due to abnormal destruction of erythrocytes by macrophage activity
31. Bilirubinuria	e. Typically due to diabetes mellitus, but may be due to stress
32. Urobilino-genuria	f. May be due to infections of the kidney or other urinary organs
33. Ketosis	g. May be due to acute inflammation of the urinary organs, kidney stones, tumors, trauma, or other kidney disease

Check Yourself

1. The four major organs of the urinary system are the kidneys, the ureters, the urinary bladder, and the urethra. **(Anatomy of the urinary system)**

2. a) Ureter

   b) Renal cortex

   c) Renal pyramid

   d) Renal medulla

   e) Renal column

   f) Renal sinus

   g) Renal capsule

   h) Renal pelvis

   i) Major calyx

   j) Minor calyx **(Anatomy of the urinary system)**

3. The kidneys are said to retroperitoneal because they are located posterior to the peritoneum of the abdominal cavity. They are located between the levels of the last thoracic and third lumbar vertebrae and are partially protected by the last two pairs of ribs. The right kidney is slightly lower due to the size of the liver superior to it on the right side of the body. The most profound layer of connective tissue surrounding the kidney is the renal capsule, a fibrous, transparent membrane continuous with the ureter. The middle layer is perirenal fat, a mass of adipose tissue that protects and anchors the kidney in place. The most superficial layer is the renal fascia, a thin layer of dense, irregular connective tissue that assists in securing the kidney to its surrounding structures. **(Anatomy of the urinary system)**

4. Blood enters the kidney through the renal artery, continuing to flow through the segmental, interlobar, arcuate, and interlobular arteries in turn. The blood then enters the efferent arterioles which supply the glomerular capillaries. The blood leaves the glomerulus in vessels named the efferent arterioles, rather than venules, because these arterioles give rise to a second capillary system, the peritubular capillaries and vasa recta. The blood leaves the nephron by the interlobular veins, arcuate veins, interlobar veins, and segmental veins to finally exit the kidney in the renal vein. **(Anatomy of the urinary system)**

5. A podocyte is a specialized epithelial cell that coves the glomerular capillaries in the renal corpuscle. Extending from each podocyte are thousands of footlike extensions called pedicels covering all the basement membrane of the capillary except for small spaces in between adjacent pedicels called filtration slits. the slits allow materials that have diffused (or filtered) through the fenestrations of the capillary to enter the Bowman's capsular space. **(Anatomy of the urinary system)**

6. a) Arcuate artery

   b) Arcuate vein

   c) Interlobular artery

   d) Interlobular vein

   e) Afferent arteriole

   f) Afferent glomerulus

   g) Bowman's capsule

   h) Efferent arteriole

   i) Proximal convoluted tubule

   j) Descending loop of Henle

   k) Loop of Henle

   l) Ascending loop of Henle

   m) Distal convoluted tuble

   n) Collecting duct

   o) Vasa recta

   p) Peitubular capillaries **(Anatomy of the urinary system)**

7. The three basic functions of a nephron are:

   a. glomerular filtration — the movement of substances across the walls of the glomerular capsules and into the renal tubule,

   b. tubular reabsorption — the return of useful substances to the blood in the peritubular capillaries and vasa recta from the renal tubule, and

   c. tubular secretion — the movement of substances from tubule cells and blood capillaries into the renal tubule. **(Physiology of the nephron)**

8. Glomerular filtration is the movement of substances across the walls of the glomerular capsule and into the renal tubule. It occurs due to the blood pressure within the glomerular capillaries that forces fluid and smaller solutes through the membranes. Glomerular filtration rate (GFR) is affected by the principal factors:

   a. the permeability of the endothelial-capsular membrane (diameter of fenestrations).

   b. surface area of the glomerular capillaries.

   c. capillary blood pressure. **(Glomerular filtration)**

9. Net filtration pressure (NFP) is the total pressure that promotes filtration. It depends upon three main pressures; glomerular blood hydrostatic pressure (GBHP), capsular hydrostatic pressure (CHP), and blood colloid osmotic pressure (BCOP). To calculate NFP, it is necessary to determine the forces that promote filtration and subtract the forces that oppose filtration.

   $$NFP = \underset{\text{promotes filtration}}{GBHP} - \underset{\text{opposes filtration}}{(CHP + BCOP)}$$

   Substituting average values for each force, a normal NFP may be calculated as:

   $$\begin{aligned} NFP &= 55 \text{ mm Hg} - (15 \text{ mm Hg} + 30 \text{ mm Hg}) \\ &= 55 \text{ mm Hg} - 45 \text{ mm Hg} \\ &= 10 \text{ mm Hg} \end{aligned}$$

   (Glomerular filtration)

10. Renal autoregulation of GFR is an intrinsic feedback system that serves to maintain an appropriate blood pressure within the glomerular capillaries. When NFP and GFR are low, the proximal convoluted tubules reabsorb more Na+, Cl-, and water than is normally reabsorbed. The macula densa cells of the juxtaglomerular complex (JGC) detect this increased reabsorption. In response, the JGC reduces its production of a vasoconstrictor resulting in the dilation of the afferent arteriole increasing the blood flow into the glomerular capillary and thus increasing the blood pressure. If the NFP and GFR become unduly elevated, the opposite effect occurs. **(Glomerular filtration)**

11. The two principal hormones affecting blood pressure are angiotensin II and atrial natriuretic peptide (ANP). Renin is released by JGC in response to decreased blood pressure in the glomerular capsule. Renin reacts with angiotensinogen (produced by the liver) in the blood, converting it to angiotensin I. In the lungs, angiotensin I reacts with angiotensin converting enzyme that completes the conversion to antiotensin II. Angiotensin II produces vasoconstriction of arterioles, stimulation of aldosterone secretion, stimulation of the thirst center of the hypothalamus, and stimulation of the ADH secretion from the pituitary gland. All of these effects serve to increase the blood pressure. ANP is secreted by cells in the atria of the heart in response to increased stretch in the atria which occurs as blood volume increases. ANP promotes diuresis and excretion of Na^+ serving to lower blood pressure. In addition, it is thought to suppress the release of ADH, aldosterone, and renin. **(Glomerular filtration)**

12. Like most blood vessels, those of the kidney are supplied by sympathetic fibers that promote vasoconstriction. When minimally activated, the sympathetic excitation produces similar vasoconstriction in both the afferent and efferent arterioles, producing little change in glomerular capillary pressure. However, during greater activation, the sympathetic fibers produce greater vasoconstriction in the afferent arterioles greatly reducing blood flow, and thus pressure, in the glomerular capillaries. **(Glomerular filtration)**

13. Tubular reabsorption is the selective recovery of fluid and solutes to the peritubular capillaries and vasa recta from the renal tubule. Most tubular reabsorption occurs in the proximal convoluted tubule (PCT). **(Tubular reabsorption)**

14. The basal face of the epithelial cells forming the walls of the PCT actively transport Na^+ from the cells into the interstitial fluid where it then diffuses into the peritubular capillaries. This is primary active transport, utilizing ATP to power the Na^+/K^+ antiporter membrane pumps. This continued removal of Na^+ from these cells maintains a lower concentration of Na^+ in the cells than will be found in the tubular filtrate, producing a simple diffusion of Na^+ across the apical face of the epithelial cells from the tubular filtrate. This transport of Na^+ promotes the passive reabsorption of water, Cl^-, and $HCO3^-$ from the tubular filtrate across the epithelial cells and eventually into the capillaries by simple diffusion. **(Tubular reabsorption)**

15. The reabsorption of nutrients such as glucose, lactic acid, amino acids, and others are accomplished by secondary active transport in the PCT. This utilizes the concentration gradient for Na^+ produced by the primary active transport of that ion. Carrier molecules called symporters act to move two substances in the same direction across a membrane. Typically this involves a co-transport of a nutrient with a Na^+ ion. Several different symporters exist that utilize the concentration gradient established by the action of the ATP activated Na^+/K^+ antiporters to recover Na^+ and a particular nutrient. **(Tubular reabsorption)**

16. Similar to the PCT, the thick ascending limb of the loop of Henle utilizes Na^+/K^+ antiporters powered by ATP to move Na^+ across the basal face of the cells. Unlike the PCT, this structure is impermeable to water. Symporter utilize the resulting Na^+ concentration gradient to co-transport Na^+, K^+, and Cl^-. The descending limb of the loop of Henle is permeable to water and approximately 15 percent of the amount of water recovered in the nephron is reabsorbed here due to osmosis. **(Tubular reabsorption)**

17. Reabsorption in the DCT and collecting ducts (CD) includes Na^+ and K^+ recovery by symporters as well as a recovery of water. By the time the tubular filtrate reaches the DCT and CD, approximately 90 percent of the filtered solutes and water have been recovered. The recovery of water to this point is called obligatory water reabsorption because the water has been obliged to follow the solutes. The remaining 10 percent of the water that reaches the DCT and CD is subject to facultative water reabsorption. The two hormones that regulate this reabsorption are aldosterone and anti-diuretic hormone (ADH). Due to the permeability variation, ion reabsorption, and the countercurrent mechanism of the nephron, the osmotic pressure in the DCT and CD is greater than that in the surrounding tissue so that water will diffuse from the DCT and CD. The amount of facultative water reabsorption is therefore largely due to the permeability of these structures to water. ADH is produced in the hypothalamus and released by the posterior pituitary gland. It stimulates the insertion of water-channel proteins into the epithelial cells of the DCT and CD, allowing for a greater reabsorption of water. Aldosterone increases the number and activity of Na^+ antiporters in the basal face of the epithelial cells of the DCT and CD, promoting water reabsorption as it follows the solutes. **(Tubular reabsorption)**

18. Tubular secretion is the movement of substances from tubule cells and blood capillaries into the renal tubule. Substance typically secreted include H^+, K^+, NH_4+, creatine, and a few drugs. Tubular secretion is largely responsible for maintaining blood pH by secretions of H^+ and eliminates wastes from the body. **(Tubular secretion)**

19. The three principal tubular secretions are H^+, K^+, and NH_3 and NH_4+. H^+ is secreted by Na^+ antiporters. In the PCT, the produced Na^+ concentration is utilized for secondary active transport with Na^+/H^+ antiporters, moving H^+ out into the tubule as Na^+ enters the epithelial cells. In the DCT, primary active transport occurs with H^+ pumps activated directly by ATP. Due to the greater permeability of K^+, almost 100 percent of the K^+ filtered is reabsorbed in the PCT, loop of Henle, and DCT. To adjust for dietary intake, DCT epithelial cells secrete a variable amount of K^+ controlled by aldosterone, plasma concentrations of K^+, and Na^+ concentration in the DCT. NH_3 is a toxic waste produced by deamination. Much of it is converted to urea. Both urea and NH_3 are eliminated in the urine. Additional NH_3 is produced in the PCT epithelial cells which quickly gains a H^+ to become NH_4+. NH_4+ can replace H^+ in Na^+/H^+ antiporters and be secreted in its place. **(Tubular secretion)**

20. The production of more concentrated urine is accomplished by the establishment of the ionic osmotic gradient in the renal medulla by the action of the thick ascending limb of the loop of Henle, an increase in DCT and CD reabsorption of water, and urea buildup in the renal medulla. **(Urine concentration)**

21. The countercurrent mechanism refers to the flow of tubular filtrate running in opposite directions in nearby parallel tubules. In other words, the tubular filtrate runs down the descending limb of the loop of Henle and then up the ascending limb.In particular, the loop of Henle of juxtamedullary nephrons descends from the renal cortex deep into the renal medulla. The descending limb is permeable to water, but relatively impermeable to solutes except urea. The ascending limb is impermeable to water, but actively transports ions into the interstitial fluid. Due to the recycling of ions and urea in the vasa recta and collecting ducts, the concentration of solutes in the interstitial fluid increases with increasing depth in the renal medulla. The overall effect works to recover water efficiently from the renal tubules. As tubular filtrate descends from the PCT into the descending limb of the loop of Henle, the interstitial fluid has a much higher concentration of solutes so that water diffuses out into the interstitial fluid producing an increasingly concentrated tubular filtrate as it descends, reaching equilibrium with the interstitial fluid at the deepest point of the loop of Henle. As the tubular filtrate begins to rise in the ascending limb of the loop of Henle, the impermeability to water and continued reabsorption of ions and other solutes produces a more dilute tubular filtrate. At the point where the tubular filtrate reaches the DCT and DC where the permeability to water is restored, once again the interstitial fluid will contain a greater concentration of solutes producing water reabsorption by osmosis. **(Urine concentration)**

22. Ureters transport urine from the renal pelvis to the urinary bladder. Peristaltic contractions of the muscular walls of the ureters propel urine toward the bladder, assisted by hydrostatic pressure and gravity. **(Ureter physiology)**

23. The ureters enter the urinary bladder at an angle. As the bladder fills, the pressure inside the bladder produces an expansion that compresses the openings to prevent a backflow of urine into the ureters. **(Ureter physiology)**

24. The expulsion of urine from the urinary bladder is called micturition (commonly called urination). Micturition is regulated by a combination of voluntary and involuntary responses. When filled, stretch receptors in the bladder walls transmit impulses to the cerebral cortex producing a conscious desire to expel urine and triggering the micturition reflex. This reflex produces parasympathetic activity which produces relaxation of the internal urethral sphincter and contractions of the detrusor muscle. However, the actual release of urine does not occur until the external urethral sphincter is relaxed. This sphincter is under voluntary control, so although a reflex activity, it can be initiated or terminated voluntarily. **(Micturition)**

25. In the male, in addition to transporting urine, the urethra serves as a conduit for semen, including sperm and the secretions of the prostate gland, seminal vesicles, bulbourethral glands, and urethral glands. **(Urethra physiology)**

26. In normal urine, the organic solutes included are urea, creatinine, uric acid, urobilinogen, and small quantities of carbohydrates, pigments, fatty acids, mucin, enzymes, and hormones. The inorganic solutes include Na^+, K^+, Cl^-, Mg^{2+}, SO_4^{2-}, $H_2PO_4^-$, HPO_4^{2-}, PO_4^{3-}, NH_4+, and Ca^{2+}. **(Urinalysis)**

27. c **(Related illnesses)**

28. e **(Related illnesses)**

29. g **(Related illnesses)**

30. f **(Related illnesses)**

31. d **(Related illnesses)**

32. b **(Related illnesses)**

33. a **(Related illnesses)**

Grade Yourself

Circle the numbers of the questions you missed, then fill in the total incorrect for each topic. If you answered more than three questions incorrectly, you need to focus on that topic. (If a topic has less than three questions and you had at least one wrong, we suggest you study that topic also. Read your textbook, a review book, or ask your teacher for help.)

Subject: The Urinary System

Topic	Question Numbers	Number Incorrect
Anatomy of the urinary system	1, 2, 3, 4, 5, 6	
Physiology of the nephron	7	
Glomerular filtration	8, 9, 10, 11, 12	
Tubular reabsorption	13, 14, 15, 16, 17, 18, 19	
Urine concentration	20, 21	
Ureter physiology	22, 23	
Micturition	24	
Urethra physiology	25	
Urinalysis	26	
Related illnesses	27, 28, 29, 30, 31, 32, 33	

Fluid, Electrolyte, and Acid-Base Homeostasis

Brief Yourself

Body fluid refers to the water and solutes contained within the body. In lean adults, water represents 55–60 percent of the body weight with approximately two-thirds of the water found within the cells and is termed intracellular fluid (ICF). The other third is extracellular fluid (ECF) which includes interstitial fluid composing about 80 percent and blood plasma composing the other 20 percent.

Body fluids contain a variety of different solutes, including both inorganic and organic substances. An inorganic substance that dissociates into ions in solution is called an electrolyte. When dissolved in water, the substance will yield positively charged cations and negatively charged anions. Acids, bases, and salts are electrolytes. In addition, a few organic molecules are ionized, such as organic acids like lactic acid or citric acid. Covalently bonded substances do not form ions upon dissolution in water and are called nonelectrolytes.

Test Yourself

1. What is meant by the fluid compartments of the body? How are they separated?
2. What are the two sources of water for the body?
3. How is fluid gain regulated?
4. How is fluid loss regulated? What hormones are involved?
5. Identify and define five methods by which the concentration of solutions may be expressed.

Questions 6–12 are matching. Match the following ion with its characteristic and function.

6. Sodium — a. Second most prevalent extracellular anion, major buffer of H^+ in plasma

7. Chloride — b. Most prevalent extracellular anion, helps balance body fluid compartments and contributes to gastric acid

8. Potassium — c. Most abundant mineral in the body, contributes to bone structure, blood clotting, neurotransmitter release, and excitability of nerve and muscle

9. Bicarbonate — d. Most abundant extracellular cation, its flow through voltage-gated channels in the plasma membrane is necessary for the generation of nerve and muscle action potentials

10. Calcium — e. Most abundant cation in intracellular fluid, plays a role in establishing resting membrane potentials and repolarizations

11. Phosphate — f. Second most abundant cation in intracellular fluid, contributes to the production of salts in bone matrix and is a cofactor for enzymes involved in the metabolism of carbohydrates, proteins, and the Na^+/K^+ ATPase for membrane pumps

12. Magnesium — g. Found forming salts to contribute to bone and teeth structure and combined with lipid, proteins, carbohydrates, nucleic acids, and ATP inside cells, as well as contributing to the buffering of the blood

13. What are the three methods by which fluid exchange occurs between the plasma and the interstitial fluid?

14. What are the mechanisms of acid-base balance?

15. What three buffer systems exist and how do they function?

16. How does the exhalation of CO_2 contribute to the maintenance of pH in the body fluids?

17. What is acidosis? What are its physiological effects?

18. What is alkalosis? What are its physiological effects?

19. How does respiratory and metabolic acidosis differ?

20. How does respiratory and metabolic alkalosis differ?

21. How can the cause of an acid-base imbalance be determined?

22. What is edema? What are some common causes of edema?

Check Yourself

1. The localization of body fluids into specific places in the body is compartmentalization. Selectively permeable membranes separate the body fluids into compartments. The plasma membrane of cells separates the intracellular and extracellular fluid compartment, just as the wall of blood vessels separates the plasma and interstitial fluid. **(Fluid compartments)**

2. The primary source of water is ingested fluids. This preformed water accounts for over 90 percent of water acquired by the body daily. The remaining water is produced as a final product of aerobic cellular respiration and is called metabolic water. **(Fluid balance)**

3. Fluid gain due to metabolism is not regulated to maintain the homeostasis of body water, thus the regulation is due to preformed water ingestion only. In other words, body fluid gain is regulated by adjustments to the amount of fluid ingested. **(Fluid balance)**

4. Fluid loss occurs in defecation, ventilation (breathing), perspiration, and urination. Under normal conditions, the majority is lost in the urine and the primary control of fluid loss is related to kidney activity. Three hormones regulate fluid loss in the kidneys; ADH, aldosterone, and ANP. ADH and aldosterone slow the fluid loss in the kidneys, while ANP increases diuresis. **(Fluid balance)**

5. The five methods of expressing concentration are:

   a. percent — numbers of grams of a substance per 100 ml of solution,

   b. millimoles per liter — number of millimoles of a substance per 1.0 liter of solution

   c. milliequivalents per liter — positive or negative charge equal to the amount of charge in 1.0 mmol/liter of H^+ (number of ions per molecule × number of charges on one ion),

   d. Milliosmole per liter — mmol/liter × number of particles per molecule upon dissolution in solution, and

   e. osmotic pressure — 1 mOsm/liter = 19.3 mm Hg. **(Concentrations of solutions)**

6. d **(Electrolytes)**

7. b **(Electrolytes)**

8. e **(Electrolytes)**

9. a **(Electrolytes)**

10. c **(Electrolytes)**

11. g **(Electrolytes)**

12. f **(Electrolytes)**

13. Substances enter and leave the capillaries in three ways; vesicular transport, simple diffusion, and bulk flow (filtration and reabsorption). Vesicular transport is due to active endocytosis and exocytosis and accounts for only a tiny fraction of plasma to interstitial fluid exchange. Most solutes move by diffusion, with bulk flow due to hydrostatic pressure providing the greatest movement of water. (**Movement of body fluids**)

14. Acid-base balance is accomplished by three mechanisms:

    a. buffering systems,

    b. exhalation of carbon dioxide, and

    c. kidney excretion of H^+. (**Acid-base balance**)

15. The three buffering systems include protein, carbonic acid-bicarbonate, and phosphate buffering systems. Protein is the most abundant buffer in the body cells and plasma. Since proteins consist of chains of amino acids, they are effect in buffering both acids and bases. The amine end of the amino acid acts as a weak base by accepting H^+, while the carboxyl group can serve as a weak acid by donating H^+.

$$NH_2 - \underset{H}{\overset{R}{\overset{|}{\underset{|}{C}}}} - COOH \Rightarrow NH_2 - \underset{H}{\overset{R}{\overset{|}{\underset{|}{C}}}} - COO^- + H^+$$

$$NH_2 - \underset{H}{\overset{R}{\overset{|}{\underset{|}{C}}}} - COO^- + H^+ \Rightarrow {}^+NH_3 - \underset{H}{\overset{R}{\overset{|}{\underset{|}{C}}}} - COOH$$

The carbonic acid-bicarbonate buffer system utilizes the bicarbonate ion as a weak base and carbonic acid as a weak acid.

H^+ hydrogen ion	+	HCO_3- bicarbonate ion	$\Rightarrow$	H_2CO_3 carbonic acid
H_2CO_3 carbonic acid	$\Rightarrow$	H^+ hydrogen ion	+	HCO_3- bicarbonate ion

Similar to the carbonic acid-bicarbonate system, the phosphate system utilizes dihydrogen phosphate as a weak acid and monohydrogen phosphate as a weak base.

OH^- hydroxide ion	+	H_2PO_4- dihydrogen phosphate	$\Rightarrow$	H_2O water	+	HPO_4^{2-} monohydrogen phosphate
H^+ hydrogen ion	+	HPO_4^{2-} monohydrogen phosphate	$\Rightarrow$	H_2PO_4- dihydrogen phosphate		

(**Acid-base balance**)

16. Concentrations of CO_2 in the blood affect the pH as demonstrated in the following reaction:

CO_2	+	H_2O	$\Rightarrow$	H_2CO_3	+	H^+	$\Rightarrow$	HCO_3^-
carbon dioxide		water		carbonic acid		hydrogen ion		bicarbonate ion

Expiration of CO_2, causing its decrease in concentration in the blood drives this react to the left, lowering the pH of the blood. **(Acid-base balance)**

17. Acidosis is the decrease of pH in the systemic arterial blood to levels below 7.35. If the pH falls below 7.0, severe depression of nervous function ensues producing disorientation and finally coma or death. **(Acid-base imbalance)**

18. Alkalosis is the increase in the pH of systemic arterial blood to levels above 7.45. In alkalosis, an overexcitability occurs in both central and peripheral nerves resulting in irritability, muscle spasm, and eventually convulsion or death. **(Acid-base Imbalance)**

19. Respiratory acidosis is due to inadequate exhalation of CO_2 which results in an increase concentration in the blood and the decrease in pH as carbonic acid is formed. Metabolic acidosis occurs as a result of ketosis or failure of the kidneys to adequately excrete H^+. **(Acid-base imbalance)**

20. Respiratory alkalosis is due to oxygen deficiency or hyperventilation causing the pCO_2 to decrease and the pH to increase as carbonic acid is lost. Metabolic alkalosis results from the intake of alkaline drugs or often as a result of a loss of HCl due to excessive vomiting. **(Acid-base imbalance)**

21. After a determination that pH is changing, and evaluation of the pCO_2 and concentration of HCO_3– is in order. If the pCO_2 is abnormal, the cause is respiratory. If the HCO_3– is abnormal, then the cause is metabolic. **(Acid-base imbalance)**

Grade Yourself

Circle the numbers of the questions you missed, then fill in the total incorrect for each topic. If you answered more than three questions incorrectly, you need to focus on that topic. (If a topic has less than three questions and you had at least one wrong, we suggest you study that topic also. Read your textbook, a review book, or ask your teacher for help.)

Subject: Fluid, Electrolyte, and Acid-Base Homeostasis

Topic	Question Numbers	Number Incorrect
Fluid compartments	1	
Fluid balance	2, 3, 4	
Concentrations of solutions	5	
Electrolytes	6, 7, 8, 9, 10, 11, 12	
Movement of body fluids	13	
Acid-base balance	14, 15, 16	
Acid-base imbalance	17, 18, 19, 20, 21	

The Reproductive Systems

21

Brief Yourself

The ability to reproduce is a characteristic of living organisms that ensures the production of new individuals of a species. In humans, the reproductive systems are dimorphic with structural and functional differences between the two sexes. Although the reproductive organs are formed during the embryonic period, they do not reach a functional maturity until puberty.

The reproductive system includes a pair of primary sex organs, the testes in males and the ovaries in females, that serve to produce gametes. The remaining structures of the reproductive systems are called the accessory reproductive organs and include ducts, glands, and external genitalia. The accessory reproductive organs serve to protect, nourish, and delivery gametes.

The male reproductive system is designed to manufacture male gametes (spermatozoa) and deliver them into the female reproductive tract and ultimately to fertilize the female gamete. The female reproductive system is designed not only to produce the female gamete (egg or ova) but also to receive the spermatozoa for fertilization. Once fertilization is accomplished, the female reproductive system assumes the greater responsibility of providing a secure and protective environment for the development of the embryo and providing for parturition.

Test Yourself

1. Label the following diagram of the structures of the male reproductive system.

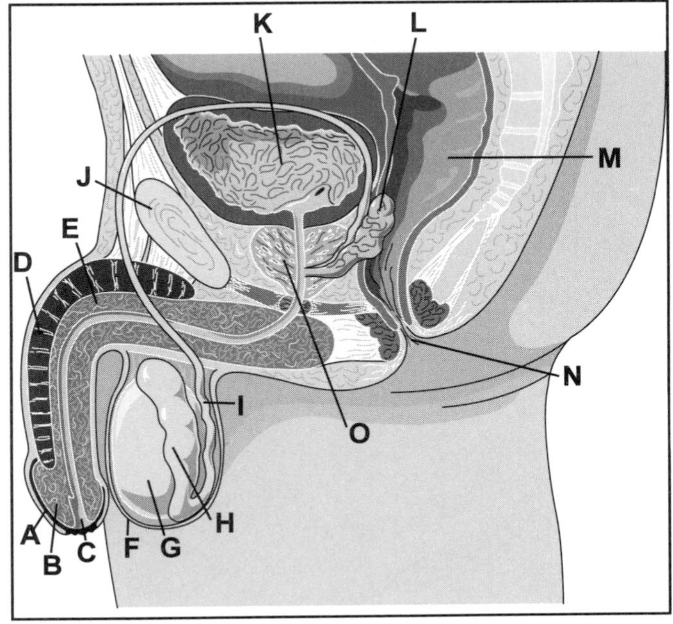

Fig. 21-1

2. Describe the structure of the scrotum. What is its function?

3. Label the following diagram of the internal structures of the testis.

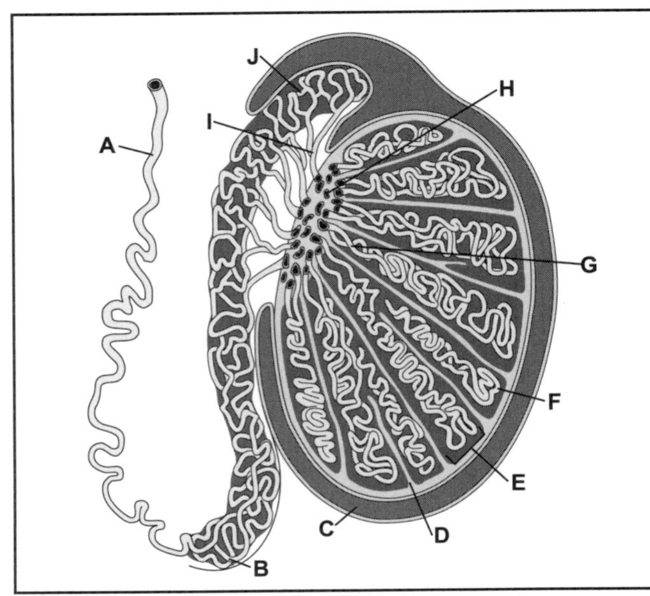

Fig. 21-2

4. Identify and describe the principal characteristics of the accessory organs forming the duct system in the male.

5. Identify the three accessory glands and their secretions in the male.

6. Label the following diagram of the structures of the penis.

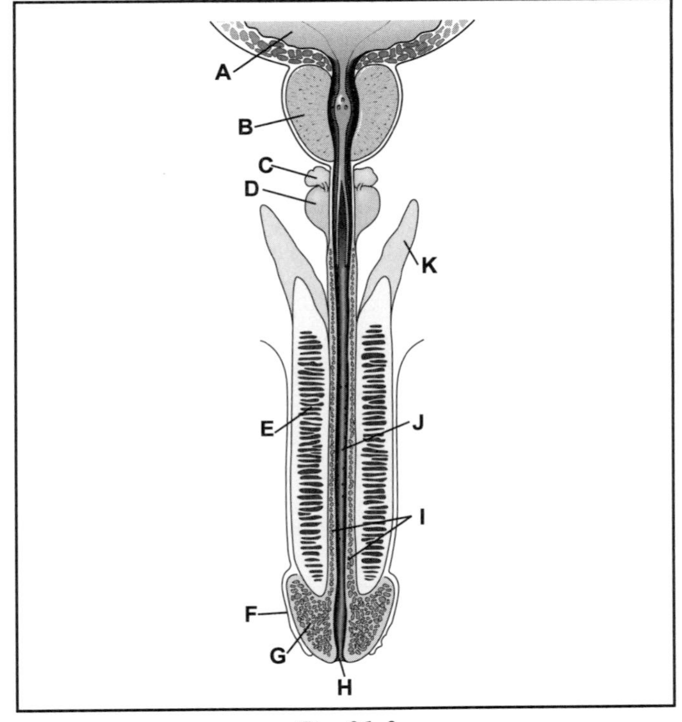

Fig. 21-3

7. What is the difference between spermatogenesis and spermiogenesis?

8. Describe the events of spermatogenesis.

9. What are the events of spermiogenesis?

10. What hormones control spermatogenesis?

11. Describe the anatomy of a spermatozoa.

12. What constitutes semen? What are the principal characteristics of semen?

13. Label the following diagram of the structures of the female reproductive system.

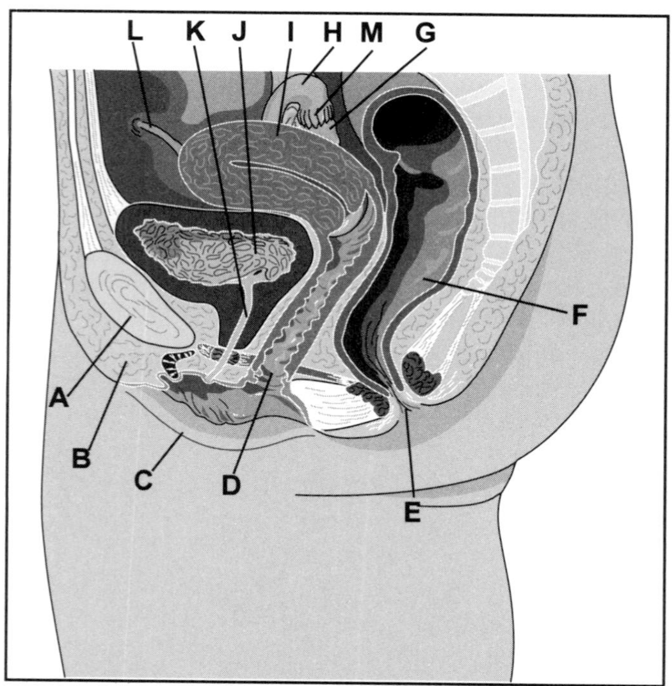

Fig. 21-4

14. What is a corpus luteum and what is its function? What is a corpus albicans?

15. Describe the events of oogenesis.

16. What is the function of the uterine tubes? How are oocytes transferred to the uterine tubes from the ovaries?

17. Describe the connective tissue structure that serve to anchor the ovary in the pelvic cavity.

18. What are the three layers of the uterus and what are their characteristics?

19. What are the components of the vulva?

20. Label the following diagram of the mammary glands.

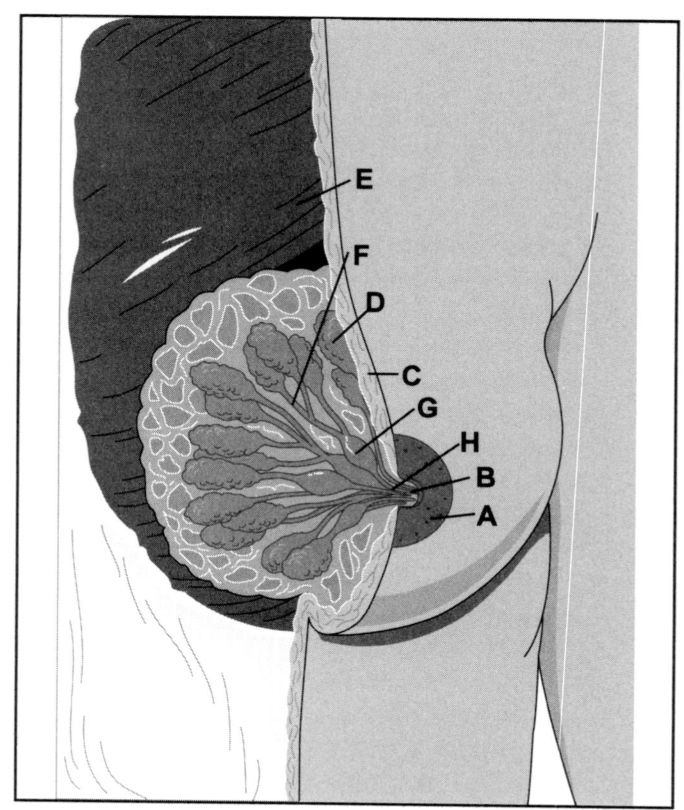

Fig. 21-5

21. What is the female reproductive cycle? What hormones regulate it?

22. What are the four phases of the female reproductive cycle and the principal events of each?

23. What are the stages of the human sexual response?

Check Yourself

1. a) Prepuce

   b) Glans penis

   c) Urethral orifice

   d) Corpus cavernosum

   e) Corpus spongiosum

   f) Scrotum

   g) Testis

   h) Epididymis

   i) Ductus deferens

   j) Pubic symphysis

   k) Urinary bladder

   l) Seminal vessicle

   m) Rectum

   n) Anus

   o) Prostate gland **(Anatomy of the male reproductive system)**

2. The scrotum is a sac of loose skin and superficial fascia that hangs from the root of the penis. It is the supporting structure of the testes. The location of the scrotum and the action of its attached muscle fibers are necessary due to the sensitivity of spermatozoa to temperature. The production and survival of spermatozoa require a temperature of 3-4° C below that of normal body temperature. The suspension of the testes in the scrotum outside the body cavities provides an appropriate thermal environment for spermatozoa. During exposure to cold temperature, the cremaster muscle elevates the scrotum to move the testes closer to the pelvic cavity for heat absorption. **(Physiology of the scrotum)**

3. a) Ductus deferens

   b) Tail of epididymis

   c) Cavity of tunica vaginalis

   d) Septum

e) Lobule

f) Seminiferous tubule

g) Tubulus rectus

h) Rete testis

i) Efferent ductule

j) Head of epididymus **(Anatomy of the male reproductive system)**

4. The duct system of the male includes:

   a. ducts of the testis — seminiferous tubules lead to straight tubules which give rise to the rete testis.

   b. epididymis — coiled epididymis ducts empty into a single ductus epididymis at the enlarged head which continually tapers to form the body and tail.

   c. ductus deferens — the less convoluted tail of the epididymis becomes the ductus deferens which ascends to penetrate the inguinal canal; its dilated end is called the ampulla.

   d. ejaculatory ducts — posterior to the urinary bladder is formed by the juncture of the ampulla and seminal vesicle.

   e. urethra — shared terminal duct of reproductive and urinary systems; sub-divided into the prostatic urethra, membranous urethra, and spongy urethra. **(Ducts of the male reproductive system)**

5. The three accessory glands of the male reproductive system are:

   a. seminal vesicles — produces alkaline, viscous fluid containing fructose, prosglandins, and clotting proteins,

   b. prostate gland — produces milky, slightly acid fluid containing citrate, acid phosphatase, and proteolytic enzymes, and

   c. bulbourethral glands — secretes an alkaline solution and mucus. **(Glands of the male reproductive system)**

6. a) Urinary bladder

   b) Prostate gland

   c) Bulbourethral gland

   d) Bulb of penis

   e) Corpus caverosum

   f) Prepuce

g) Glans penis

h) Urethral orifice

i) Corpus spongiosum

j) Spongy urethra

k) Crus of penis **(Anatomy of the male reproductive system)**

7. Spermatogenesis is the process by which the seminiferous tubules produce haploid spermatozoa. Spermiogenesis is the final stage of spermatogenesis in which spermatids mature into motile spermatozoa. **(Spermatogenesis)**

8. Spermatogenesis begins with the mitotic division of spermatogonium (stem cells) producing a migration of the daughter cells away from the basement membrane of the seminiferous tubule. This daughter cell differentiates to form primary spermatocytes which undergo meiotic divisions. Meiosis I (reduction division) occurs as a diploid (2n) primary spermatocyte replicates its DNA to form 46 chromosomes (each made of two identical chromatids). The chromosomes then associate with their homologous pairs so that there are 23 pairs of duplicated chromosomes. At this point, recombinations (crossing-over) can occur. The first meiotic division pulls one duplicated chromosome from each pair toward the opposite pole followed by cytokinesis to produce two secondary spermatocytes. The random assortment of chromosomes provides additional genetic variation at this point. Meiosis II occurs without further duplication of DNA such that chromosomes are separated into individual chromatids followed by cytokinesis to produce four haploid (n) spermatids. **(Spermatogenesis)**

9. Spermiogenesis is the maturation of spermatids into motile spermatozoa. It begins with the activity of the Golgi apparatus to package acrosomal enzymes and the positioning of the acrosome at the anterior end of the nucleus and the centrioles at the opposite end. Elaboration of microtubules follows to form the flagellum as mitochondria multiply around the proximal end of the flagellum. In the last step, excess cytoplasm is reduced and the mature spermatozoa are released. **(Spermatogenesis)**

10. Spermatogenesis is ultimately controlled by gonadotropin releasing hormone (GnRH) from the hypothalamus. Its release stimulates the release of luteinizing hormone (LH) and follicle-stimulating hormone (FSH) from the pituitary gland. LH stimulates the production of testosterone which will act synergistically with FSH to stimulate sustenacular cells in the testes. These cells release androgen-binding protein (ABP) which maintains a high concentration of testosterone in the seminiferous tubules stimulating spermatogenesis. **(Spermatogenesis)**

11. A mature spermatozoon consists of a head region containing the acrosome and nucleus, a midpiece containing mitochondria, and a tail (flagellum). **(Spermatogenesis)**

12. Semen is a mixture of mature spermatozoa and seminal fluid, which is the fluid component of semen. Seminal fluid is a mixture of the secretions of seminal vesicles, prostate gland, and bulbourethral glands. A normal ejaculate has a volume of 2.5–5.0 ml with a sperm count of between 50 and 150 million spermatozoa per milliliter. Semen is slightly acidic and milky in appearance due to the secretions of the prostate. It is sticky due to bulbourethral and seminal vesicle secretions including clotting agents, fructose, and antibiotics. **(Semen)**

13. a) Pubic symphysis

    b) Mons pubis

    c) Labium

    d) Vagina

    e) Anus

    f) Rectum

    g) Ovary

    h) Uterine tube

    i) Uterus

    j) Urinary bladder

    k) Urethra

    l) Round ligament

    m) Fimbriae **(Anatomy of the female reproductive system)**

14. The corpus luteum is the remnants of an ovulated mature follicle. It produces progesterone, estrogen, relaxin, and inhibin until such time as it degenerates. The degenerated corpus luteum becomes a corpus albicans, composed primarily of fibrous tissue. **(Anatomy of the female reproductive system)**

15. Oogenesis is the formation of haploid (n) secondary oocytes in the ovary. During early fetal development, germ cells differentiate into diploid (2n) oogonia. A few develop into primary oocytes by beginning meiosis I, but the process is suspended until puberty. At birth, between 200,000 and 400,000 oogonia and primary oocytes remain in each ovary. Most will degenerate due to atresia, but about 400 will mature and ovulate during a woman's reproductive lifetime. Each primary oocyte is surrounded by a single layer of follicular cells producing a primordial follicle. These mature into primary follicles by the addition of first one and then additional layers of cuboidal follicular cells. The primary follicle continues to grow, developing into a secondary follicle. With the influence of puberty, each month one secondary follicle resumes the suspended meiosis I producing haploid cells of unequal size. The large cell is the secondary oocyte and the smaller the polar body. The polar body may or may not divide again, but the secondary oocyte suspends its meiosis II until such time as fertilization occurs. The secondary oocytes will be released by the ovulation of the mature follicle. If fertilization occurs following ovulation, the secondary oocyte will complete meiosis II, again unequal cytokinesis producing a polar body prior to the combination of chromatids from the spermatozoa. **(Oogenesis)**

16. The uterine tubes transport secondary oocytes and fertilized ova from the ovaries to the uterus. Unlike male ducts, the uterine tubes are not truly continuous with the ovaries. Finger-like projections of the uterine tube called fimbriae drape over the ovary. Ovulation casts a secondary oocyte into the peritoneal cavity, where some are lost. The fimbriae become active at ovulation, sweeping the surface of the ovary.

The ciliated cells of the fimbriae beat to draw the oocyte in and the rhythmic beating of cilia and peristaltic contractions move the oocyte toward the uterus. **(Ducts of the female reproductive system)**

17. Each ovary is held in place by the ovarian ligament which anchors it medially to the uterus. In addition, the suspensory ligament and mesovarium form a part of the broad ligament that serves to anchorthe ovary to the pelvic wall. **(Anatomy of the female reproductive system)**

18. The three layers of the uterus are:

    a. perimetrium — part of the visceral peritoneum; consists of simple squamous epithelium and areolar connective tissue,

    b. myometrium — consists of three layers of smooth muscle, and

    c. endometrium — highly vascularized and velvety layer; innermost consists of simple columnar epithelium with an underlying endometrial stroma of areolar connective tissue and glands. **(Anatomy of the female reproductive system)**

19. The vulva is the external genitalia of the female and consists of:

    a. mons pubis.

    b. labia majora.

    c. labia minora.

    d. clitoris.

    e. vestibule. **(Anatomy of the female reproductive system)**

20. a) Areola

    b) Nipple

    c) Adipose tissue

    d) Lobule (Alveoli)

    e) Pectoralis major

    f) Mammary duct

    g) Lactiferous sinus

    h) Lactiferous duct **(Anatomy of the breast)**

21. The female reproductive cycle is the cyclical changes that occur in the ovaries and uterus of nonpregnant females. Each cycle takes approximately one month and encompasses ovarian and uterine cycles. The cycle is ultimately controlled by GnRH from the hypothalamus that stimulates the production of LH and FSH. FSH stimulates the secretion of estrogen by growing follicles. LH stimulates the further

development of ovarian follicles, promotes formation of corpus luteum, brings about ovulation, and stimulates production of estrogen, progesterone, relaxin, and inhibin. **(Female reproductive cycle)**

22. The four phases of the female reproductive cycle are:

    a. menstrual phase — in the ovary small secondary follicles begin to enlarge; in the uterus blood, tissue fluids, mucus, and epithelial cells are discharged due to ischemia following the constriction of the uterine spiral arteries,

    b. preovulatory phase — in the ovary secondary follicles continue to develop secreting estrogen until one becomes dominant; estrogen and inhibin secreted by the dominant follicle decrease FSH release and smaller follicles are inhibited; in the uterus estrogens stimulate repair of the endometrium,

    c. ovulation phase — a mature follicle ruptures releasing a secondary oocyte into the pelvic cavity; the mature follicle collapses to become a corpus hemmorrhagicum eventually developing into the corpus luteum, and

    d. postovulatory phase — in the ovary the corpus luteum persists if the oocyte is fertilized, maintained by human chorionic gonadotropin (hCG) from the chorion of the embryo; with no fertilization the corpus luteum degenerates into the corpus albicans; in the uterus the progesterone and estrogens produced by the corpus luteum promote growth; glands secrete glycogen, vascularization increases, endometrium thickens. **(Female reproductive cycle)**

23. The human sexual response includes excitement (arousal) which produces vasocongestion of blood in the genital tissues. Secretion of lubricating fluids begins, primarily in the female, as well as increases in heart rate and blood pressure, breathing rate, and skeletal muscle tone. These changes are sustained in the plateau stage which may last for many minutes. During the plateau stage most females, and many males, exhibit a rashlike redness of the head and chest called a sex flush. Orgasm follows during which the male ejaculates and both male and female experience rhythmic muscle contractions accompanied by intense pleasurable sensations. Males enter a refractory period following ejaculation during which time a second orgasm is impossible. Females may experience two or more orgasms in rapid succession. The final stage is resolution characterized by relaxation in the genital tissues, heart rate, blood pressure, breathing rate, and muscle tone. **(The human sexual response)**

Grade Yourself

Circle the numbers of the questions you missed, then fill in the total incorrect for each topic. If you answered more than three questions incorrectly, you need to focus on that topic. (If a topic has less than three questions and you had at least one wrong, we suggest you study that topic also. Read your textbook, a review book, or ask your teacher for help.)

Subject: The Reproductive Systems

Topic	Question Numbers	Number Incorrect
Anatomy of the male reproductive system	1, 3, 6	
Physiology of the scrotum	2	
Ducts of the male reproductive system	4	
Glands of the male reproductive system	5	
Spermatogenesis	7, 8, 9, 10, 11	
Semen	12	
Anatomy of the female reproductive system	13, 14, 17, 18, 19	
Oogenesis	15	
Ducts of the female reproductive system	16	
Anatomy of the breast	20	
Female reproductive cycle	21, 22	
The human sexual response	23	

Development and Inheritance

22

Brief Yourself

Developmental anatomy is the study of the progression of growth and maturation from the fertilization of a secondary ooctye to the formation of an adult. The development from a single cell (fertilized zygote) to an adult containing approximately 50 trillion cells occurs in two broad stages; prenatal and postnatal. Although postnatal development is impressive, the degree of growth from a single cell to a full term fetus during prenatal development is truly dramatic. From fertilization to birth, the sequence of events of prenatal development includes fertilization, implantation, gestation, labor, and parturition.

The development and growth of an individual are guided by the codes held in the gene-bearing chromosomes received from the sperm and egg of the parents. Inheritance is the passage of hereditary traits from one generation to the next. This process provides the acquisition of characteristics from the parents to the offspring.

Test Yourself

1. Describe the events of fertilization.
2. What is a cleavage? How often do they occur?
3. How does a morulla and a blastocyst differ?
4. Label the following diagram of implantation.

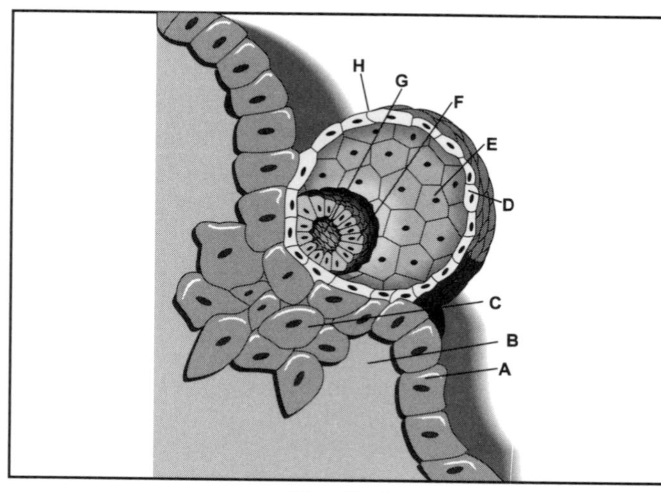

Fig. 22-1

5. How does the blastocyst merge and burrow into the endometrium?

6. What is the embryonic period? What is the fetal period?

7. What is gastrulation?

8. Identify the embryonic membranes and their functions.

9. Label the following diagram of an embryo.

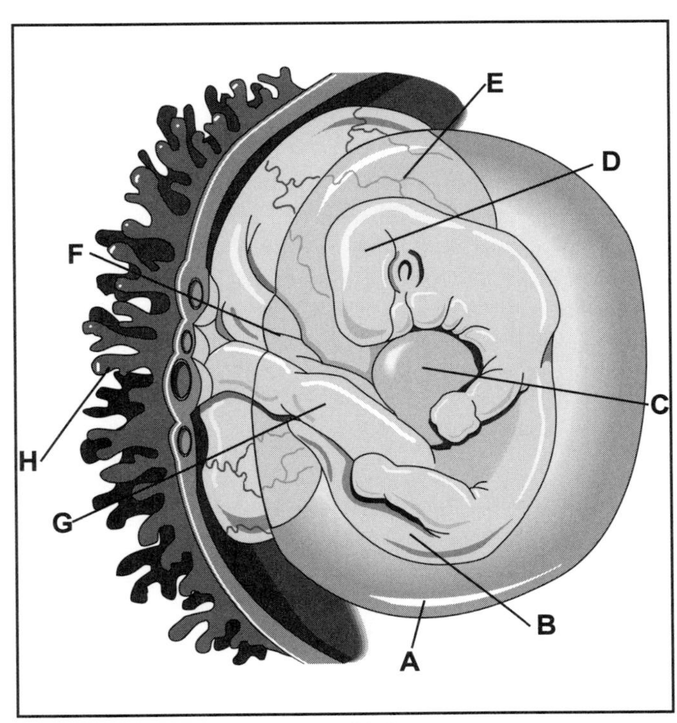

Fig. 22-2

10. What is organogenesis? What is the first major event of organogenesis?

11. What is an umbilical cord? Describe the vascular connection between mother and fetus.

Questions 12–20 are matching. Match the following month of fetal growth with its characteristics.

12.	First month	a.	Head and body more proportionate, skin wrinkled, capable of survival, assumes coronal position, in male testes descend
13.	Second month	b.	Additional subcutaneous fat accumulates, lanugo shed, nails extend to fingertips
14.	Third month	c.	Eyes, nose, ears not yet visible, vertebral column and canal form, limb buds form, heart forms and begins to beat, CNS appear
15.	Fourth month	d.	Head becomes less disproportionate, eyelids separate and eyelashes form, substantial weight gain, skin wrinkled, type II alveolar cells begin to secrete surfactant
16.	Fifth month	e.	Eyes far apart, eyelids fused, nose flat, ossification begins, limbs become distinct, major blood vessels form
17.	Sixth month	f.	Eyes almost fully developed, eyelids still fused, nose forms bridge, external ears present, ossification continues, limbs fully formed and nails develop, heartbeat is detectable, urine starts to form, fetal movement that cannot be detected by mother
18.	Seventh month	g.	Subcutaneous fat deposited, skin less wrinkled, excellent chances of survival
19.	Eighth month	h.	Head large in proportion to the body, face assumes human features, hair on head, many bones ossified, joints form
20.	Ninth month	i.	Head less disproportionate to body, lanugo covers body, brown fat forms, quickening occurs

21. What hormones help to regulate pregnancy and what are their roles? Which is utilized for early pregnancy tests?

22. What are the stages of labor?

23. What regulates lactation?

Questions 24–33 are matching. Match the following term with its definition:

24.	Allele	a.	The physical expression of a gene
25.	Mutation	b.	Containing the same alleles on homologous chromosomes
26.	Phenyl-ketonuria	c.	Alternate forms of a gene on the same locus of a chromosome
27.	Genotypes	d.	Chart to determine possible haploid gamete crosses to form diploid fertilized eggs
28.	dominant trait	e.	a permanent heritable change in a gene
29.	Recessive trait	f.	Inability to manufacture phenylalanine hydroxylase
30.	Homozygous	g.	Genetic makeup of an individual
31.	Heterozygous	h.	An allele that masks the effect of another allele
32.	Phenotype	i.	Containing different alleles on homologous chromosomes
33.	Punnett square	j.	An allele whose presence is completely masked by another allele

34. What is incomplete dominance?

35. What is multiple-allele inheritance?

36. What is a sex-linked trait? Why do males exhibit recessive sex-linked traits more commonly than females?

37. What principal environmental factors influence the expression of genes?

Check Yourself

1. Fertilization occurs when the genetic material of a sperm merges with that of an egg to form a zygote. Although only one sperm performs the actual fertilization, the action of a number of sperm is required to allow the entry of the fertilizing sperm. Approximately 1 percent of the sperm reach the egg, typically in the uterine tube. The capacitation of the membrane around the acrosomes of sperm encountering the egg produces a release of enzymes that penetrate the corona radiata and zona pellucida. This typically allows only one sperm to enter the egg, triggering the completion of meiosis II in the secondary oocyte and leading to fertilization. In addition, the entry of a sperm produces a cortical reaction preventing the entry of other sperm (polyspermy). **(Fertilization)**

2. After fertilization, rapid mitotic divisions of the fertilized zygote occur called cleavages. The first cleavage occurs approximately 24 hours after fertilization with each succeeding cleavage occurring sooner. By the end of the third day, there are 16 cells. **(Fertilization)**

3. A morulla is a solid mass of blastomeres that is produced in the first few days following fertilization. Blastomeres are the progressively smaller cells produced by cleavages. At approximately five days, the blastomeres have developed into a hollow ball of cells that enter the uterus called a blastocyst. **(Fertilization)**

4. a) Simple columnar epithelium of uterus

   b) Endometrial stroma

   c) Syncytiotrophoblast

   d) Trophoblast

   e) Blastocele

   f) Inner cell mass

   g) Amniotic cavity

   h) Blastocyst **(Implantation)**

5. As the blastocysts adheres to the wall of the uterus, it orients so that its inner cell mass is toward the endometrium. Two layers develop in the outer cell layer (trophoblasts) of the blastocyst, the syncytiotrophoblast and cytotrophoblast. The syncytiotrophoblasts secrete enzymes that digest and liquefy the endometrial cells allowing the blastocyst to become buried in the endometrium. **(Implantation)**

6. The embryonic period refers to the first two months of development during which time the developing human is called an embryo. The remaining months of development until parturition are called the fetal period, during which time the developing human is called a fetus. **(Embryonic Development)**

7. Gastrulation refers to the conversion of the inner cell mass of a blastocyst into the primary germ layers of ectoderm, endoderm, and mesoderm. **(Embryonic Development)**

8. The embryonic membranes include:

   a. yolk sac — serves to nourish embryo in early stage of development, early site of blood formation, produces primitive germ cells.

   b. amnion — surrounds embryo creating a cavity filled with amniotic fluid that protects the developing fetus, regulates fetal temperature, protects skin of fetus, and stores fetal wastes.

   c. chorion — becomes the principal embryonic portion of the placenta for exchange of materials between the fetus and the mother.

   d. allantois — small, vascularized portion of yolk sac for early blood production, serves to form umbilical connection between the mother and fetus. **(Embryonic development)**

9. a) Amniotic cavity

   b) Tail of embryo

   c) Heart of embryo

   d) Head of embryo

   e) Yolk sac

   f) Allantois

   g) Body stalk (umbilicus)

   h) Chorionic villi **(Embryonic development)**

10. Organogenesis is the formation of body organs (and organ systems) that develop following gastrulation. The first major event of organogenesis is neurulation, the development of the brain and spinal cord. **(Embryonic development)**

11. The umbilical cord is a vascular connection between the mother and fetus. It consists of two umbilical arteries, one umbilical vein, and supporting mucus connective tissue (Wharton's jelly). The chorion (fetal portion of the placenta) produces fingerlike projections into the decidua basalis (maternal portion of the placenta). These projections, chorionic villi, are filled with fetal blood capillaries that exchange fluid with the pooled blood in the intervillous space in the decidua basalis. Maternal arterioles deliver oxygenated and nutrient rich blood to the intervillous space and maternal venules carry deoxygenated and waste rich blood away. Since the maternal and fetal circulatory systems are discontinuous, the placenta also serves as a protective barrier since many microorganisms cannot cross it. **(Placenta and umbilical cord)**

12. c **(Fetal growth)**

13. e **(Fetal growth)**

14. f **(Fetal growth)**

15. h **(Fetal growth)**

16. i **(Fetal growth)**

17. d **(Fetal growth)**

18. a **(Fetal growth)**

19. g **(Fetal growth)**

20. b **(Fetal growth)**

21. Pregnancy is regulated primarily by progesterone and estrogen. Early in pregnancy, the chorion produces hCG which mimics the effect of LH and maintains the activity of the corpus luteum so that it continues to produce estrogen and progesterone. The presence of hCG serves as the basis for early pregnancy tests. By the third or fourth week of pregnancy the chorion begins to secrete estrogen and by the sixth week secretes progesterone as well. These secretions maintain the pregnancy. hCG will decline during the fourth and fifth week and the corpus luteum will degenerate. The placenta also produces relaxin that increases the flexibility of the pubic symphysis and other ligamentous or cartilagenous structures. The chorion of the placenta also produces human chorionid somatomammotropin (hCS) that serves to prepare mammary glands for lactation. **(Hormone of pregnancy)**

22. Labor occurs in three stages:

    a. dilation stage—from labor onset to complete dilation of cervix.

    b. expulsion stage—from complete cervical dilation through parturition.

    c. placental stage—from parturition until afterbirth expulsion. **(Labor)**

23. Lactation is regulated primarily by prolactin which is released in response to a prolactin releasing hormone (PRH) secreted by the hypothalamus. Its activity is inhibited by high levels of progesterone produced by the placenta. Following parturition, the decline in progesterone allows the prolactin to produce milk secretion. The suckling action of the infant initiates nerve impulses that stimulate the release of oxytocin from the pituitary gland. Oxytocin produces a contraction of the myoepithelial cells of the mammary glands that serves to eject milk. **(Lactation)**

24. c **(Genotype and phenotype)**

25. e **(Genotype and phenotype)**

26. f **(Genotype and phenotype)**

27. g **(Genotype and phenotype)**

28. h **(Genotype and phenotype)**

29. j **(Genotype and phenotype)**

30. b **(Genotype and phenotype)**

31. i **(Genotype and phenotype)**

32. a **(Genotype and phenotype)**

33. d **(Genotype and phenotype)**

34. Incomplete dominance occurs when neither member of an allele pair is dominant over the other. The heterozygote will result in a phenotypic expression that is intermediate between the homozygous dominant or homozygous recessive expression. For example, in humans the texture of hair is produced by incomplete dominance. The homozygous dominant has curly hair, the homozygous recessive has straight hair, and the heterozygous has wavy hair. **(Genetic dominance)**

35. Multiple-allele inheritance occurs when genes have more than two alternate forms. An individual will receive only two of the possible alleles. For example, the ABO blood grouping in humans is based on multiple-allele inheritance. There are three different alleles, I^A, I^B, and i. I^A and I^B are inherited as dominant alleles over i, but neither exhibit dominance over each other. Thus, possible genotypical and phenotypical expressions are:

    I^A I^A — type A blood

    I^A i — type A blood

    I^B I^B — type B blood

    I^B i — type B blood

    I^A I^B — type AB blood

    i i — type O blood **(Genetic dominance)**

36. Sex-linked traits are nonsexual traits those that exist on the X chromosomes but are absent on the Y chromosome. The inequalities in the expression of sex-linked traits between males and females are due to the recessive sex-linked traits. In order for a female to express a recessive trait, both X chromosomes must carry it (homozygous recessive). Since males have no matching allele that could repress expression, the recessive traits are expressed in males as often as the dominant. **(Genetic dominance)**

37. A teratogen is any agent or influence that can cause developmental defects in the fetus. The principal environmental factors that affect gene expression of the fetus are:

    a. chemicals and drugs — many chemicals can cross the placenta that are dangerous to the fetus.

    b. cigarette smoking — tars and nicotines of smoke are teratogens, also lowers blood oxygen content of the mother.

    c. irradiation — ionizing irradiation is a potent teratogen. **(Genes and the environment)**

Grade Yourself

Circle the numbers of the questions you missed, then fill in the total incorrect for each topic. If you answered more than three questions incorrectly, you need to focus on that topic. (If a topic has less than three questions and you had at least one wrong, we suggest you study that topic also. Read your textbook, a review book, or ask your teacher for help.)

Subject: Development and Inheritance

Topic	Question Numbers	Number Incorrect
Fertilization	1, 2, 3	
Implantation	4, 5	
Embryonic development	6, 7, 8, 9, 10	
Placenta and umbilical cord	11	
Fetal growth	12, 13, 14, 15, 16, 17, 18, 19, 20	
Hormone of pregnancy	21	
Labor	22	
Lactation	23	
Genotype and phenotype	24, 25, 26, 27, 28, 29, 30, 31, 32, 33	
Genetic dominance	34, 35, 36	
Genes and the environment	37	

Katharine
Dexter